AF593090

Positive Pressure Attack for Ventilation & Firefighting

Positive Pressure Attack for Ventilation & Firefighting

Kriss Garcia, Reinhard Kauffmann, and Raymond Schelble

Disclaimer

The recommendations, advice, descriptions, and methods in this book are presented solely for educational purposes. The authors and publisher assume no liability whatsoever for any loss or damage that results from the use of any of the material in this book. Use of the material in this book is solely at the risk of the user.

PennWell Corporation
1421 South Sheridan Road
Tulsa, Oklahoma 74112-6600 USA
800.752.9764
+1.918.831.9421
sales@pennwell.com
www.pennwellbooks.com
www.pennwell.com
www.fireengineering.com

Director: Mary McGee
Managing Editor: Jerry Naylis
Production/Operations Manager: Traci Huntsman
Senior Design Editor: Robin Remaley
Production Editor: Tony Quinn
Cover Designer: Clark Bell

Library of Congress Cataloging-in-Publication Data

Garcia, Kriss.
Positive pressure attack for ventilation & firefighting / Kriss Garcia, Reinhard Kauffmann, and Raymond Schelble.
p. cm.
Includes bibliographical references.
ISBN-13: 978-1-59370-048-5 (hardcover)
ISBN-10: 1-59370-048-2 (hardcover)
1. Fire extinction--Technique. 2. Displacement ventilation. I. Kauffmann, Reinhard. II. Schelble, Raymond. III. Title.
TH9339.G37 2006
628.9'25--dc22

2006018091

Printed in the United States of America

4 5 6 12 11

To our committed brother and sister firefighters worldwide who continually search for a better way

Contents

Acknowledgments xiii

Introduction xv

I Overview of Fireground Ventilation 1

1 **What Is Effective Ventilation? 3**
Products of Combustion 3
Defining Fireground Ventilation 5
Systematic 6
Coordinated 7
What Is Effective Ventilation? 7
Benefits of Effective Ventilation 9
1. Assists with survival and rescue of trapped victims 10
2. Protects firefighters 13
3. Aids firefighter entry 16
4. Rapid advance to the seat of the fire 17
5. Decreases fire spread 18
6. Decreases property damage 19
References 20

2 **Conventional Ventilation Methods 21**
Horizontal Ventilation 22
Vertical Ventilation 22
Historical background 22
Basic operations 24
Mechanically Assisted Ventilation 26
Negative pressure ventilation 26
Positive pressure ventilation 27
Water Fog Ventilation 28
References 29

3 Limitations of Conventional Methods 31
Limitations of Horizontal Ventilation 31
Limitations of Vertical Ventilation 32
The time factor 33
Firefighters in harm's way 34
How much do power tools help? 35
Three main ventilation challenges 36
Everything has changed but ventilation 37
Changes in New Construction Methods and Materials 38
Wood 38
Steel 40
Plastics 43
Training and Education Issues 46
Educating citizens 47
Flawed firefighter training 48
Limitations of Other Ventilation Methods 49
Horizontal ventilation 50
Mechanically assisted ventilation 50
Water fog 54
Additional Factors to Consider 54
Who ventilates? 54
Tactical considerations 56
Increased BTU evolvement 57
Difficulty with roof coverings 58
Cannot push down ceiling 58
Compartmentalized construction 61
Power tools may not work 62
Firefighters may fall into the fire 63
Sprinkler system activation 64
Ventilation by Building Construction Type 65
Type I: Fire resistive 65
Type II: Noncombustible 66
Type III: Ordinary 69
Type IV: Heavy timber or mill construction 73
Type V: Wood frame 75
Aspects of Indirect Attack 78
An Alternative 78
References 79

II Positive Pressure Attack **81**

4 Understanding Positive Pressure Ventilation 83
The Origins of PPV 83
Turning the fans around 84
Building a better blower 85

PPV Fundamentals. 86
How PPV works 86
Anatomy of a blower 89
Positive pressure terminology. 92
Setting up PPV at the Fire Scene 93
The ventilation opening 93
The exhaust opening 93
Safety and precautions 95
Fine-tuning PPV. 96
Blower Configurations. 97
Single blower 98
In-line (series) blowers 98
Parallel blowers 99
Combination blowers. 100
V-point blowers 101
Stacked blowers 102
Structural Influences 102
Areas with limited access 102
Basements 103
Multiple-story dwellings 104
Commercial buildings 105
Unintended exhaust openings. 108
No exhaust opening. 109
Controlled Ventilation 109
References 110

5 PPV + Initial Fire Attack = PPA = Coordinated Attack 111
Taking PPV a Step Further. 112
The next logical question 113
The beginnings of PPA 113
A fatal house fire 114
The pieces come together 115
PPA Fundamentals. 117
Positioning the blower 117
Creating an exhaust opening 119
Beginning pressurization and fire attack. 122
Fine-tuning PPA. 125
An effective exhaust opening 125
Coordination with attack. 127
Keys to successful PPA. 129
Why PPA Is Dynamic. 131
An Exciting New Method 135

III Beyond the Basics. 139

6 Precautions for Using PPV and PPA 141

The Importance of Training. 142

Exhaust Point Safety 144

Belching flame and smoke 144

Protecting firefighters. 146

Victims at the exhaust point. 148

Exposure hazards. 149

Make Sure the Fire Is Out 151

Blowers and rekindles 152

Turn off the blower 153

Problem Situations. 154

Backdraft conditions 154

Combustible dust and flammable vapors 159

Smoke explosion potential. 162

Carbon monoxide alarms. 162

Adjusting for weather. 163

7 PPA and Beyond 165

Effective Uses for Positive Pressure and Blowers 165

Vehicle fires 166

Dumpster fires 168

Aircraft fires. 169

Chimney fires. 171

Confined space rescue 172

Sequential ventilation. 175

Large-scale ventilation 177

Pressurizing for Exposure Protection 178

Exposure types. 180

Pressurizing exposures 183

Pressurization by design. 185

Pressurizing specific exposure types 186

Ventilating High-Rise Buildings 193

Factors influencing smoke movement. 193

Tactical considerations. 200

8 Introducing PPA to the Department 209

The SLCFD Experience 209

Fluttering toilet paper 210

The next step 211

Going inside. 211

"Test driving" in buildings 213

Acceptance evolves. 215

Salt Lake City's Teaching Methods 217

The intellectual side 217

The emotional side. 218

9 Positive Pressure Evolutions **221**
Ensuring Competency 221
PPA Evolutions 224
PPA with fast attack. 226
PPA with water supply. 228
Discussion 230
Training issues 231
Training benefits 233
References 234

10 A Deeper Understanding **235**
Twenty Burning Questions about Positive Pressure Attack. 236
1. Will blowing fresh air into a structure involved in fire push fire throughout the building? 236
2. Will PPA push fire throughout voids in the building? 238
3. Will PPA spread fire through voids in a balloon frame–constructed building? 241
4. Why does PPA sometimes make the fire appear to intensify at the exhaust point? 247
5. Does PPA require additional staffing to accomplish the fire attack? 249
6. Does PPA delay rescue and fire control?. 250
7. What if firefighters suspect there may be trapped victims between the fire and exhaust opening?. 252
8. What is the best location for the ventilation opening and exhaust opening?. 255
9. At what point in the operation should the blower begin pressurizing the structure? 258
10. Is it dangerous to put the blower too close to the seat of the fire? 260
11. Can a crew make an exhaust opening in the wrong location? 262
12. What about the CO introduced by the gasoline-powered blowers?. 265
13. When should firefighters use PPA?. 268
14. What about using PPA to fight attic fires? 269
15. What about using PPA to fight basement fires? 272
16. What if a room, basement, or building has only one opening? 273
17. What are some potential problems with PPA? 276
18. Under what conditions is PPA not recommended? 279
19. How can PPA protect exposures?. 282
20. What size of building can be ventilated by a blower?. 284
Conclusion 284
Benefits of PPA. 285
References 285

11 Final Thoughts **287**

Appendix A: Why Positive Pressure Works **289**
The Science behind Positive Pressure 289
Avogadro's Hypothesis. 290
Boyle's Law 290
Charles' Law 291
Dalton's Law 293
The Ideal Gas Law. 293
Gas laws in real life 294

Appendix B: Salt Lake City Fire Department PPA Evolutions **295**
Ventilation Evolution #1 295
Ventilation Evolution #2 296

Appendix C: AMCA Standards for Blowers **297**
References 300

Appendix D: Acronyms and Abbreviations **301**

Bibliography **303**

Index **307**

Acknowledgments

For more than a decade, we have interacted with many dedicated fire service professionals from around the world regarding tactical operations. Without their commitment to firefighter safety, and their sincere interest in improving the fire service, this book would not have been possible.

We are reluctant to mention any single individual, department, or agency for fear of leaving some out. There are so many. However, we would also be remiss if we did not make an attempt. The following is a list of some of the major contributors to the knowledge and content in this book, in no particular order: the administration and firefighters of the Salt Lake City Fire Department and Tooele City Volunteer Fire Department; Cliff Allmon; Dexter Coffman; Martha Ellis; Richard Moseley; Jack Tidrow; Rowe Harrison; John Mittendorf; Phil McLaughlin; Gary Seidel; Mark Yates; Mark Egan; Tempest Technology, Inc.; Dr. Rich Gordon; Air Movement & Control Association International; Super Vacuum Manufacturing Company, Inc.; Rick and Jill Black; the National Fire Protection Association; National Fire Academy; the National Institute of Standards and Technology; the Institution of Fire Engineers; the more than 6,000 students who have challenged and encouraged us; PennWell, for giving us the opportunity to get our message down on paper; and, finally, our families, who supported us these last four years while we wrote and researched and discussed.

To these people and everyone who helped and supported us along the way, we offer a sincere "Thank you."

Introduction

The workplace of firefighters has been undergoing a dramatic transformation. Slowly, steadily over decades, the materials used to build and furnish homes, commercial buildings, warehouses, high-rises, and just about everything else have been changing. More and more mass has been shaved from structural materials. Plastics have replaced natural materials for furniture and decorating. As a result, buildings do not withstand fire for long, and fires burn far hotter and faster than in the past, producing far more deadly gases.

For decades, leaders in the fire service like Francis L. Brannigan and John Mittendorf have been educating firefighters on early collapse hazards and poisonous products of combustion. Victims trapped in these buildings are at risk from burning synthetic materials that very quickly create lethal conditions.

The faster that heat and products of combustion can be removed from a burning building interior, the better for everyone. Ideally, ventilation started early and coordinated with fire attack provides the ultimate benefit for firefighters searching for victims and attacking the fire. It also helps unprotected victims survive until rescuers can find them. This is a difficult ideal to accomplish, so most firefighters have learned to work as well as we can in limited visibility until the interior eventually clears.

This book presents a way we can make ventilation work, and how and when it can do the most good. Positive pressure attack (PPA) provides safe, effective ventilation that can be accomplished in time to benefit firefighters and victims. PPA works in new buildings and old, can be applied to a wide variety of situations, and does not require special staffing or expensive equipment.

We decided to write this book because there is confusion and mis-information within the fire service about PPA, and there are many firefighters who do not have access to accurate information. We are firefighters. Since we started this project, our intent has been to write a practical guide that would benefit the broad spectrum of the fire service—from the firefighters who position the blowers to the incident commander who directs the operation, and from the urban career firefighter to the rural volunteer. To this end, we have avoided technical jargon and have tried to keep the ideas and language as straightforward as possible.

The book is divided into three sections. The first discusses the ideal of ventilation coordinated with fire attack and examines traditional methods against this ideal. The second section is the nuts and bolts: positive pressure ventilation (PPV), PPA, and precautions that will make positive pressure a safe choice for firefighters and the public. The third section deals with how to adapt positive pressure to specific situations, discusses practical training and education issues, and answers 20 of the most common questions we hear regarding PPV and PPA.

If you are already convinced of the benefits of positive pressure, this book can put you on a path to better understand how it works and how to make it work more effectively. If you are skeptical, please read this text with an open mind. Of course, part of this book represents our opinions, but we have tried hard to explain the basis for those opinions as objectively as possible.

We sincerely hope this book helps you to gain a better understanding of positive pressure and PPA, and becomes a valued reference as you continue to develop your skills and knowledge.

I

Overview of Fireground Ventilation

To get a clear picture of where positive pressure attack fits into fire service ventilation practices, it is necessary to examine the various conventional methods in use. Information in the three chapters of section I includes:

- A description of, definition of, and reasons for ventilation
- A basic overview of ventilation methods
- The established criteria that determine effective ventilation
- A discussion of how conventional methods measure up to the criteria for effective ventilation

1
What Is Effective Ventilation?

Products of combustion can range in harmfulness from irritating to deadly depending on their chemistry and concentrations. Removing these products and heat is what ventilation is all about.

This chapter begins with a general discussion on what these by-products of combustion include. Later, the difference between ventilation and *effective* ventilation is delineated.

Products of Combustion

All firefighters begin learning about the by-products generated by fire—heat, smoke, and products of combustion—in their first, most basic classes on firefighting. Heat and smoke in an environment can interfere with or completely prevent firefighters from entering and working, and can make it impossible for victims to survive (fig. 1–1).

Products of combustion can include heat and smoke as well as thousands of chemicals. These chemicals can cover a spectrum ranging from irritants to those that can kill outright, and include some that cause serious health problems later. Through study, observation, and experience, firefighters learn to greatly respect the hazards posed by heat, smoke, and toxic products of combustion. For the purpose of this book, unless otherwise modified, the term *products of combustion* will refer to heat, smoke, and other toxic products of combustion.

Fig. 1–1 Heat and smoke can interfere with or completely prevent firefighters from entering a structure and can make it impossible for victims to survive. Courtesy Ray Schelble

The following list contains some common toxic substances that have been found in products of combustion in structure fires. It is not by any means complete.

- **Carbon monoxide (CO).** The most abundant of the toxic gases, it is usually found in greater concentrations in fires that are not well ventilated.
- **Hydrogen cyanide (HCN).** It is the most toxic of gases (20 times more toxic than carbon monoxide). Hydrogen cyanide is derived from burning materials containing nitrogen, such as wool, silk, nylons, plastics, and polyurethanes.
- **Carbon dioxide (CO_2).** This is a gas produced in large volumes during fires. It causes respiratory rate increase, thus increasing the inhalation of other toxic gases.
- **Acrolein (C_3H_4O).** This is a sensory and pulmonary irritant produced by the smoldering of all cellulose materials. It has been shown to cause severe lung disorders hours after exposure.
- **Hydrogen chloride (HCl).** This is commonly derived from polyvinyl chloride (PVC). It is a sensory and pulmonary irritant released when items that contain chlorine are involved in fire.

Products of combustion confined in a structure create serious problems for firefighters and trapped victims alike. In structures, products of combustion rise until they are blocked by a ceiling or roof and then begin to fill the space downward, a process referred to as *mushrooming*. Because mushrooming starts at the highest point in a space, the atmosphere near the floor will be the last to become contaminated. Given enough time, though, the products of combustion become the atmosphere in the entire space.

The effects of products of combustion in a structure are varied, and none could be considered good. Effects include:

- Breathing products of combustion can result in critical or long-term health problems or death.
- The lethal atmosphere may not support life at all.
- Vision is obscured.
- Temperatures and flammable gases can build to explosive levels.
- Structure and contents can sustain damage far away from any flame.

Defining Fireground Ventilation

The practice of removing products of combustion from inside a building, or *ventilation*, has long been an important fireground tactic. Over the years, firefighters have developed and refined fireground ventilation methods to remove products of combustion from structures.

The definition of ventilation as it relates to firefighting has been covered in virtually every fire service textbook, trade journal, and training manual. The definition may vary a bit in the exact wording, but the overall meaning is always the same. By general definition, it is the systematic and coordinated removal of heat, smoke, and products of combustion to replace them with cooler, fresher air. Two words in this definition are essential to understanding what good ventilation requires: *systematic* and *coordinated*.

Systematic

Webster's Third New International Dictionary defines the term *systematic* as "following or observing a plan."[1] In the case of ventilation, the word *systematic* refers to practiced and proven methods used with the fire attack to address certain conditions (namely, the presence of products of combustion).

As with just about all operations on the fireground, good ventilation does not happen by chance. It is a planned, practiced operation that follows a system.

For example, an incident commander (IC) who has practiced the incident command system (ICS) and properly takes command of an incident is proceeding methodically. This IC is following a plan that outlines a way to successfully accomplish management of the incident. Likewise, effective ventilation proceeds methodically and follows a plan that outlines its implementation (fig. 1–2). If done correctly, systematic ventilation should accomplish favorable, predictable results.

Fig. 1–2 Just as the incident commander proceeds methodically and follows a plan to manage an incident, ventilation must also proceed methodically and follow a plan to achieve consistent, predictable results. Courtesy Martha Ellis

Coordinated

"Ventilation timing is extremely important and must be carefully coordinated with rescue and fire attack activities," according to Alan V. Brunacini.[2] The term *coordinated* is a key concept when assessing the value of any type of ventilation.

Coordination, again according to Webster's, means to "act together in a smooth, concerted way."[3] To be most effective, ventilation and interior attack crews should work together in a "smooth, concerted way." Without coordination between ventilation and fire attack, neither will be as successful. For this reason, ventilation and fire attack should be considered equally important.

As one will see in the next chapter, coordinated ventilation has been extremely difficult to achieve. Coordinating ventilation with fire attack is a key concept and will be referred to often.

What Is Effective Ventilation?

Removing products of combustion at any time during an incident provides some benefit. However, the greatest benefit from ventilation comes when it supports interior crews in performing the all-important tasks of rescue and extinguishing the fire. In fire situations, where speed and efficiency are of the greatest importance, this requires that ventilation, rescue, and fire attack be accomplished as soon as possible and in coordination with each other (fig. 1–3).

In *Structural Firefighting,* the authors sum up the importance of ventilation:

- "Ventilation is one of the incident commander's most important considerations."
- "Modern methods involve a coordinated effort in which ventilation is given a high priority."
- "Proper ventilation can have a positive effect on the three fireground priorities of life safety, extinguishment, and property conservation."[4]

Fig. 1–3 Speed and efficiency are of the greatest importance in fire situations. In most cases, ventilation should be accomplished immediately. Courtesy Ray Schelble

Brunacini makes it clear that ventilation should be coordinated with fire attack. He also has a strong opinion on when it should begin. "Ideally, [ventilation] should occur just ahead of interior crews advancing hose lines," he says.[5]

To be considered effective, ventilation should be started before attack crews advance on the fire. This does not mean crews are attempting to ventilate while search and rescue as well as fire attack are being made. Rather, it means that ventilation must be accomplished *before* the fire attack and search and rescue operations begin.

Having a working environment that is clear enough for firefighters to see and cool enough to work in standing up allows interior crews to make a rapid advance on the main body of fire. A tenable atmosphere also allows firefighters to rapidly search large areas by sight and remove victims who have not spent long times exposed to toxic levels of products of combustion.

Without coordinated ventilation, firefighters are forced to crawl through deadly interior environments trying to find victims on hands and knees, and locating fire by crawling close to the area that is most dangerous. If the primary goal is to mitigate the incident as quickly and safely as possible, then failing to launch a coordinated attack actually works against this goal.

Timing is important. Only ventilation started early in the fire attack can help interior crews that battle the clock to search the building and attack the fire. The best ventilation job in the world falls short if it comes too late.

But there is more to it than just timing. "Fire conditions and ventilation are highly interrelated, and they must be balanced," Brunacini writes.[6] All the good timing in the world will be for nothing if the ventilation is not adequate to clear the atmosphere in the structure. Whatever the method of ventilation, it must be appropriate to deal with the size of the fire and the volume of products of combustion in the building.

Therefore, to meet the criteria for effective ventilation, it must begin before fire attack and before search and rescue crews enter. Also, it must be balanced with fire conditions so the inside of the structure will be clear for interior crews to do their work.

Benefits of Effective Ventilation

The concepts of ventilation are not new, and performing ventilation is certainly not without hazards. The "Salt Lake City Fire Department Orientation Manual" (1950 edition) opens a discussion on ventilation with an important statement (fig. 1–4). It says that

> *ventilation used properly is a great advantage to firemen, it will permit entering and working in an area that otherwise would be unbearable. But if not used wisely, it can be the cause of losing a whole building or perhaps causing a conflagration and consequent heavy damage or loss of life.*[7]

Fig. 1–4 The 1950 edition of the "Salt Lake City Fire Department Orientation Manual" states, "Ventilation used properly is a great advantage to firemen, it will permit entering and working in an area that otherwise would be unbearable." Courtesy Salt Lake City Fire Department (SLCFD)

Effective ventilation provides six important benefits. Nothing in this list has dramatically changed in the past 200 years.

1. Assists with survival and rescue of trapped victims

Toxic gases, decreasing oxygen for victims to breathe, and lack of visibility inside the structure are the main reasons why firefighters ventilate. It becomes obvious in terms of victim survival.

In a structure fire, hazards such as burns and structural collapse definitely pose a threat to victims. But hazards from products of combustion unquestionably have the biggest impact on victim survival. The National Fire Protection Association's (NFPA) *Fire Protection Handbook* states that "most building-related fire deaths are directly related to these products of combustion. Death often results from oxygen deprivation in the bloodstream, caused by the replacement of oxygen in the blood hemoglobin by carbon monoxide."[8]

This is not the only problem that products of combustion pose to victim survival. The *Fire Protection Handbook* also states, "Dense smoke obscures visibility and irritates the eyes and can cause anxiety and emotional shock to occupants."[9] This leads to panic and disorientation, which makes escape far more difficult.

A study undertaken by the Fire Protection Research Foundation of NFPA and reported in *Fire Chief Magazine* has examined the impacts of smoke and heat on victims from a different angle. "For most of the history of fire science and fire safety, our efforts have focused on how much smoke would kill a person," says the foundation's president, Rick Mulhaupt. "Now, we're recognizing that many people die in fires—not because smoke killed them on the spot—but because smoke or heat prevented them from getting out of the building."[10]

Nationally, fires in dwellings account for a large number of fatalities each year. The NFPA reports, "Most fire fatalities in one- and two-family dwellings are a result of smoke inhalation, not burns, and most victims die in a room that is not the room of fire origin. In other words, they die as a result of a fire that has produced enough smoke to spread beyond the room in which the fire began."[11]

Fire development. Test fire research was conducted by the Southwest Research Institute in a "furnished, simulated hotel room with attached corridor and remote room." This research demonstrated that after a chair made the transition from the smoldering phase to flaming ignition, "the fire progressed rapidly to flashover within about 8 min., accompanied by the development of life-threatening conditions of temperature, CO, HCN, and O_2 depletion in the room of origin." After flashover, conditions in the remote room also deteriorated to the point that "test animals (rats) . . . became incapacitated within about 2 min. after flashover, with death occurring from CO asphyxiation at approximately 11 min. . . . It is likely that humans in the remote room would have been incapacitated and killed in approximately the same time intervals."[12]

It is important to consider what must happen in the eight minutes it takes to get to the flashover stage for there to be a possibility for a successful rescue. Somebody must notify the fire department, firefighters must turn out and respond to the scene, and fire hoses must be stretched. Firefighters must enter the building and then find and remove the victims.

The whole time victims are awaiting rescue, they are inhaling toxic products of combustion, and the temperature of these inhaled products is quickly rising to lethal levels. In an autopsy, these victims have black soot in their airways and lungs, along with lethal levels of CO in the blood stream.

Effects of heat and water. The National Research Council of Canada has demonstrated that when air in excess of 300°F is inhaled, victim survival can be expected only for very short periods of time. If moisture is present in the heated atmosphere, survival of victims cannot be expected at all.[13] During the initial stages of fire development, structure fires rapidly exceed the 300°F threshold temperature of survival, with the floor being the last survivable space. Coupled with the introduction of moisture from a well-intentioned attack line, victims will not survive long in this environment without ventilation.

Victims trapped in burning structures are also subject to burns on their skin from the high temperatures. At temperatures above 130°F, skin is subject to burns within seconds. As with inhalation, the time it takes for skin to burn decreases when water vapor is present. This is a sure thing when attack lines are advanced and water converts to steam and is pushed to the floor level, where victims are awaiting rescue. Because of the systemic impacts of thermal burns, victims are subject to shock, hyperthermia, and fatal heart attacks due to increased heart rate and inhalation of lethal toxins and heat.

Thermal imaging equipment has become popular in the fire service since the early 1990s. Granted, these tools can help locate victims and lead to faster rescue times. But if the purpose of these tools is to locate victims, unless the interior atmosphere is cool and uncontaminated enough for rescuers to make entry without personal protective equipment (PPE), the victims inside will not have much of a chance at survival.

The bottom line is that attack lines alone will not increase victim survival time. In fact, introducing water to a heated environment produces steam, which will actually decrease survival time. (This topic is developed further later in the text.) In informal surveys conducted by the authors in the course of teaching classes across the country and internationally, several thousand firefighters were surveyed. They were asked if they had ever been involved in a rescue where a victim in the fire area survived after the application of water to an interior fire. Less than 2% of the reported victims had been discharged from the hospital and survived 30 days.

The importance of sight. According to the *Fire Protection Handbook*, "The effect of smoke obscuration on lethality is not direct, nonetheless, it is real in that escape can be hindered or precluded where visibility does not exist."[14] According to *Company Officer*, "For civilians who may still be present in the building, ventilation can easily make a difference in their survival. Ventilation improves the likelihood that civilians will be found during search and rescue operations and that they will be savable."[15] Without ventilation, the victims continue to inhale lethal products of combustion and suffer the effects of heat until they eventually get rescued or they die. Inhalation of carbon monoxide and other poisonous gases in products of combustion also triggers human behaviors that adversely affect motor skills and impair judgment.

Anything that speeds up the search benefits the survivability of victims. Because of modern protective gear, firefighters can quickly move deeper into a burning structure than ever before. But smoke decreases the ability to search effectively and efficiently. Ventilation allows firefighters to use vision in addition to touch and hearing to make a rescue (fig. 1–5). The ability to see may make the difference between search and rescue that results in a survivable victim or a body recovery. If the atmosphere is so contaminated that firefighters cannot see a victim, the chance of a victim surviving even seconds in that environment is very poor.

Effective ventilation also allows firefighters to make a rapid exit with a found victim because they can actually see the safest and quickest way out. Having the ability to see the nearest exit is often the only way that victims (and firefighters) may self-rescue.

Fig. 1–5 Effective ventilation implemented early in the incident allows firefighters to use vision in addition to touch and hearing during fire attack and rescue. Courtesy Kriss Garcia

2. Protects firefighters

Most firefighters can easily remember their first search-and-rescue training exercise in a simulated fire structure (fig. 1–6). Their memories likely conjure up the experience of having their sight obscured in some fashion and then crawling along the floor with their arms and legs outstretched in hopes of being able to touch something that might—just might—be a victim.

Too many news reports and trade journal articles relate accounts of trapped firefighters who died in unsuccessful attempts to find a way out of a smoke-filled structure. Sometimes these casualties were only a few feet from the nearest exit when they died, but heavy smoke did not allow them to find their way.

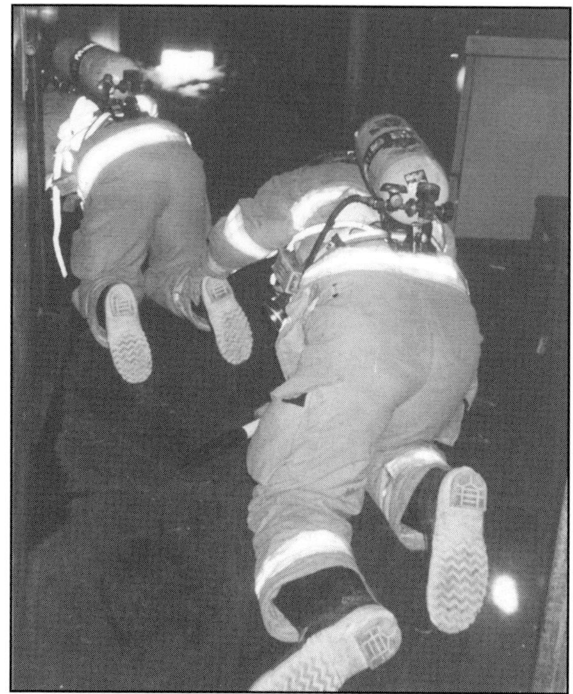

Fig. 1–6 The first search-and-rescue training for new firefighters usually consists of blocking vision through the face mask and having them crawl along the floor trying to touch something that might be a victim. Courtesy Ray Schelble

In a report examining conditions leading up to this tragedy, the LAFD stated,

> *The best thing you can do for the situation is to aggressively ventilate the structure. It must be coordinated with attack crews so as not to allow the fire to possibly progress towards the area where people are lost/trapped . . . In the areas where we lost smoke, crews were able to extricate the victims much faster.*[16]

Impaired vision due to the presence of heavy smoke is one of the undisputed hazards encountered by firefighters. Dense smoke can hide all sorts of traps, including shafts, open stairwells, burned-out floors, exposed wires, and other hazards.

When firefighters work inside a building, their turnouts protect them. Their gear buys them precious time while awaiting rescue if they become trapped or injured. Eventually, though, they run out of air, and despite all their training and education, they may pull off the face piece of their self-contained breathing apparatus (SCBA). At this moment they suffer the same respiratory doom as the victims they seek because of the smoke and heat—unless the area is ventilated.

Without question, safety is greatly enhanced when firefighters can see their surroundings (fig. 1–7). The answer is ventilation, but it must be performed in support of the fire attack, and it must be accomplished early enough to be effective in benefiting the crews performing the primary search.

Fig. 1–7 Although firefighters working inside buildings are protected by their turnouts and SCBA, safety is greatly enhanced when they can see their surroundings. Courtesy Kriss Garcia

With effective ventilation in place during their training exercise mentioned earlier, the firefighters would have been able to better see where they were going and what their outstretched arms and legs were touching. One can imagine how much more successful they would have been in finding the victim.

Modern protective gear and SCBA can protect firefighters from extremely inhospitable environments. Ventilation can save them a lot of punishment.

3. Aids firefighter entry

Firefighters are familiar with the situation of responding to a structure fire, with heavy black smoke billowing from the doors, windows, and attic. More often than not, they initially made entry into a wall of black, acrid smoke and ceiling temperatures that may have exceeded 600°F. Although they could not see, they still advanced the hose line and felt along the walls and floor to guide them deeper into that dark, hot abyss. Their progress was slow until the ventilation crew opened up a hole in the roof that started to channel that toxic atmosphere outside and replaced it with fresh air. On seeing the glow, they opened the nozzle on the fire. It knocked down the fire and cooled the interior a bit, which produced steam. This made the products of combustion less buoyant and slowed down the rate at which those products could rise through the ventilation hole in the roof. Despite this, the firefighters could make more headway because they gained an advantage with the heat and smoke reduction. They could begin to see, even if it was only a short distance ahead at first.

To be most effective, ventilation must be in place and working during the time firefighters make entry (fig. 1–8). As mentioned previously, effective ventilation must be coordinated with the initial hose line attack.

Fig. 1–8 To be effective, ventilation must be in place and working during the time firefighters make entry. Courtesy Kriss Garcia

4. Rapid advance to the seat of the fire

A well-known firefighting axiom states that in most cases, firefighters should not open nozzles on smoke alone when advancing to the seat of the fire. Withholding water until fire is visible is almost universally recognized as the most effective way to attack a fire.

But how long does it take to see the fire if the attack crew is battling heavy smoke and heat? When vision is limited, interior crews must choose the most probable route to where they believe the fire is located. They grope their way through dark hallways and rooms until they sense an increase in heat. Hopefully they will eventually be able to see the telltale orange glow that identifies the fire. This inflicts physical punishment on the firefighters, which slows their advancement to the seat of the fire and reduces the time they can work. It also subjects working firefighters to a flammable atmosphere.

Effective ventilation early in a fire protects attack crews from intense fire conditions. It also assists in rapid discovery of the fire and quick knockdown (fig. 1–9). A timely and well-placed ventilation operation must be strategically designed to direct the heat and smoke away from the attack crews and limit the fire to its area of origin. Only then can a truly rapid attack be made.

Fig. 1–9 Effective ventilation early in a fire protects attack crews from intense fire conditions and assists in rapid discovery of the fire and a quick knockdown. Courtesy Kriss Garcia

5. Decreases fire spread

Left alone, a structure fire spreads vertically and horizontally, always seeking fuel and being drawn to lower atmospheric pressure. Ventilating the structure does two things to disrupt the spread of the fire.

First, ventilation removes heat. As heat is removed, it slows the ignition of potential fuels. For every decrease of 18°F, the speed of the chemical reaction leading to combustibility decreases 50%.[17]

Second, proper ventilation directs the fire along a path that can be determined by fire crews, perhaps through an opening where it has self-vented or out an opening created by firefighters. For example, when the roof is opened and the ceiling breached, the increased interior pressures caused by heating forces the products of combustion to the decreased atmospheric pressure outside through the opening. The natural tendency of heat to rise also contributes to this type of ventilation. The fire naturally follows the path of least resistance through the opening firefighters have provided.

However, it is seldom that simple. When firefighters make a ventilation hole in a roof to vent the building's livable space, the fire has now been introduced into an attic space that may have been previously free of fire. If decreasing the spread of fire is a reason for ventilation, then the potential hazards from introduction of fire into previously uninvolved spaces must be assessed, carefully monitored, and controlled for any extension (fig. 1–10).

Fig. 1–10 Potential hazards from introduction of fire into a previously uninvolved attic space must be assessed, carefully monitored, and controlled for any extension. Courtesy Tooele City Volunteer Fire Department (TCVFD)

Another complication comes from the extreme amounts of heat (measured in British thermal units, or BTUs) released by the synthetic materials that fuel today's structure fires. "Few people have any concept of the heavy fire load and high rate of fire growth of today's furniture," according to the National Fire Protection Agency (NFPA).[18] The operational norm of a 4-foot (ft) by 4-ft hole for vertical ventilation is seriously inadequate. An inadequate ventilation hole risks spreading a contained fire through the attic area because the BTUs cannot escape fast enough. There will be more discussion on this later.

6. Decreases property damage

In a fire, smoke leaves sooty residue on personal belongings and stains the interior of the structure walls and windows. Heat melts and distorts plastics, interior fixtures, electronics, and other items.

Property conservation is an important concern of firefighters and benefits the building owner and tenants when it comes to repairs, displacing residents, and loss of income. Even a small fire can produce smoke that permeates the entire structure, including closets, upper floors, and attic spaces. Effective ventilation goes a long way toward decreasing the damage caused by smoke and fire (fig. 1–11).

Fig. 1–11 Effective ventilation goes a long way toward decreasing the damage caused by smoke and fire. Courtesy Kriss Garcia

Effective ventilation when the fire is still in the incipient or free, burning stage reduces smoke and heat damage. Water damage will also be lessened because firefighters can find the seat of the fire sooner. For example, one should consider that a 2½-inch (in.) attack line with a 1¼-in. tip will deliver 325 gallons per minute (gpm), which calculates to approximately 1 ton of water per minute. In multistory buildings, the damage caused by large volumes of water may be substantial on floors below the fire.

The next chapter looks at the basics of the various conventional methods of ventilation that are accepted and in use by the U.S. fire service.

References

1. *Webster's Third New International Dictionary of the English Language, Unabridged*. 1993. Springfield, MA: Merriam-Webster.
2. Brunacini, Alan V. 1985. *Fire Command*. Quincy, MA: National Firefighters Protection Association.
3. Guralnik, David B., editor. 2003. *Webster's Dictionary*. New York, NY: Warner Books.
4. Klaene, Bernard J., and Russell E. Sanders. 2000. *Structural Fire Fighting*. Quincy, MA: National Firefighters Protection Association.
5. Brunacini. 1985.
6. Ibid.
7. Salt Lake City Fire Department. 1950. *Salt Lake City Fire Department Orientation Manual*. Salt Lake City, UT: Salt Lake City Fire Department.
8. National Fire Protection Association. 2003. *Fire Protection Handbook*. 19th ed. Quincy, MA: National Fire Protection Association.
9. Ibid.
10. Ibid.
11. Ibid.
12. Ibid.
13. Ibid.
14. Ibid.
15. Smoke, Clinton H. 1999. *Company Officer*. Albany, NY: Delmar Publishers.
16. Los Angeles City Fire Department. 1998. *Lowery Fire RIT Exercise*. Los Angeles, CA: Los Angeles City Fire Department.
17. Campbell, Doug. 1998. *The Campbell Prediction System*. 2nd ed. Ojai, CA: Ojai Printing and Publishing Co.
18. National Fire Protection Association. 2003.

2

Conventional Ventilation Methods

NFPA 1001, Standard for Fire Fighter Professional Qualifications, outlines general performance and knowledge standards for firefighters.[1] A key performance issue in the guidelines on ventilation concerns when "the structure is cleared of smoke." In the description it also calls for ladder skills as a way to provide this smoke-free environment. Although ladder and vertical ventilation skills are certainly necessary for a firefighter, by no means are they necessary to provide a smoke-free interior environment (fig. 2–1).

Fig. 2–1 Although ladder and vertical ventilation skills are certainly necessary for a firefighter, they are not often necessary to provide a smoke-free interior environment, especially in a residence or small commercial building. Courtesy SLCFD

The information in this chapter is deliberately basic. It is intended to provide general descriptions and a foundation for discussing current and accepted methods of ventilating a structure that is on fire. Conventional ventilation methods fit into four general categories: horizontal, vertical, mechanically assisted, and water fog.

Horizontal Ventilation

Horizontal ventilation is often the simplest and quickest means of improving the interior environment for fire attack. As the fire burns and creates pressure, it follows the path of least resistance, which often occurs naturally as windows fail. Crews can often enhance the effect by breaking out windows near the fire area.

Tactically speaking, horizontal ventilation is the use of doors and windows to ventilate across a floor of a building. Ideally, windows or doors on the leeward side of the building are opened first and, if possible, at the top of the window opening. The next step is to open the windward side of building at a lower level, if possible. Horizontal ventilation is often the quickest means of beginning some level of ventilation and does not require much equipment.

Vertical Ventilation

The historic ideal of vertical ventilation is simple: cutting an opening above the fire allows trapped, hot gases to escape from the structure. It directs the fire to burn freely upward and out of the building (fig. 2–2). In theory, this will limit mushrooming and the horizontal spread of the fire and also makes the interior more tenable for the victims and firefighters.

Historical background

Vertical ventilation has been around for years. Benjamin Franklin, one of the fathers of the U.S. fire service, was also reported to be a vertical ventilation tactician who recognized the value of removing products of combustion from the fire building. As the story goes, he made the connection from observing fire burn in his fireplace as heat and smoke channeled up the chimney.

Fig. 2–2 The historic ideal of vertical ventilation is simple: cutting an opening in the roof above the fire directs the fire and products of combustion to burn freely upward and out of the building. Courtesy SLCFD

In vertical ventilation a hole is cut (a chimney of sorts) in the roof or at a high point on a building. Since the heated products of combustion rise, eventually some of them escape through the hole. This same basic concept has been used essentially the same way for more than two centuries (fig. 2–3).

Fig. 2–3 The same basic concept of vertical ventilation has been used for more than two centuries. Courtesy SLCFD

Also around Ben Franklin's time, hook-and-ladder companies came about. The job of the hook-and-ladder crew was to create a ventilation opening for the fire by actually hooking the roof and pulling parts of it off. While these initial holes were meant to remove the combustible thatch, they also created a hole in the roof, and a form of vertical ventilation was realized. Most of these early efforts, although based on sound logic, produced little in the way of effective results, except that eventually these holes did allow some heat and smoke to escape.

Basic operations

Vertical ventilation requires crews to work on the roof of a building and demands that crews pay close attention to roof conditions throughout the operation. Making a vent hole can be as simple as breaking out skylights, but it most often involves cutting a hole in the roof (fig. 2–4).

Fig. 2–4 Making a hole for vertical ventilation can be as simple as opening a roof structure, but most often involves cutting a hole. Courtesy SLCFD

Starting a vertical ventilation operation requires a minimum of two ladders—one for access and another for emergency egress. Procedures in many fire departments and recommendations in some firefighting texts also call for a charged hoseline on the roof for "protection."

Crews cutting a vent hole on a roof are faced with two major decisions: where to cut the hole and how big to make it. As for the first major decision, the ideal location of the hole is directly above the fire or as close to directly above the fire as possible. This is called a primary vent hole. Often the location of the primary vent hole requires crews to act on a best guess of where the fire is located.

Another use of vertical ventilation is to stop a fire from mushrooming under the roof. This type of vent hole is called a direction hole. Trenching (cutting a continuous hole completely across a roof to stop fire spread underneath the roof) is an example of how a direction hole can be used. A trenching operation is three-dimensional, meaning that a roof trench should also have a corresponding trench in the ceiling below. It is this area that is protected by a hoseline from spreading to the uninvolved portion beyond the trench.

The second major decision is how big to cut the hole. Which guidelines to use in deciding how big to make a primary vent hole has been the subject of much discussion in the fire service. At one time the standard was to start with a 4-ft by 4-ft hole and enlarge it in steps of 2 ft at a time until the smoke stopped venting under pressure. Another rule of thumb dictated a vent hole be 10% of the size of the building area affected by the fire. As with the location of the vent hole, the size to cut the hole can also be an inexact science.

Cutting the hole requires using axes or power saws to cut through the roof deck. Crews must be cautious to not cut through roof supports that would compromise the structural integrity of the roof. This is important to the safety of the crews, since the assault of the fire on roof supports can accelerate weakening of the roof structure. After the hole is cut, ceilings and other structures must be cleared from underneath to create a clear opening into the fire area.

One factor to be considered in vertical ventilation is that when it is working, the attack stream works against it. As the interior environment cools with the application of water, the products of combustion will not rise as well as they did when the interior was hot. High levels of CO will remain in the structure.

Mechanically Assisted Ventilation

Mechanical ventilation methods are variations of horizontal ventilation, which, in its basic form, involves creating openings on the windward and leeward sides of a building to permit wind to blow through the structure. Under the right circumstances, it was found to be a fairly effective way to clear products of combustion after the fire was out.

The first real technological advancement in ventilation and the biggest advance since vertical ventilation was the smoke ejector. A smoke ejector is an electric-powered fan used to move products of combustion from a space at about 5,000 cubic feet per minute (cfm). Mechanical ventilation includes two methods: negative pressure ventilation and positive pressure ventilation.

Negative pressure ventilation

For negative pressure ventilation, smoke ejectors are placed in an exterior opening, usually a door, with the air flow directed outside (fig. 2–5). This encourages air movement from inside to the outside, which helps clear out the interior atmosphere of the building. Creating an opening on the other side of the area to be cleared and using wind to advantage will help encourage air movement. Since products of combustion tend to rise, this type of ventilation works better if the smoke ejector is positioned as near to the ceiling as possible. A variety of attachments and accessories are available to suspend the smoke ejectors from doors, ladders, or door or window casings.

Fig. 2–5 In negative pressure ventilation, a fan is placed inside a building in an opening to exhaust the interior atmosphere. Courtesy SLCFD

Positive pressure ventilation

At some point, firefighters performing negative pressure ventilation discovered that they could greatly enhance the effects of the wind by turning around the smoke ejectors and positioning them on the windward side of the structure to blow into a door opening. Crews then found if they moved the smoke ejector outside the door and several feet away, it would force air through the entire door and provide several benefits. First, the smoke ejector no longer created an obstruction in the doorway, which allowed more freedom for crews, hoselines, and equipment to move in and out. Second, the blower increased the pressure in adjacent areas inside the structure. This provided far more efficient airflow throughout the structure than with negative pressure or horizontal ventilation, and could clear products of combustion in a relatively shorter period of time (fig. 2–6).

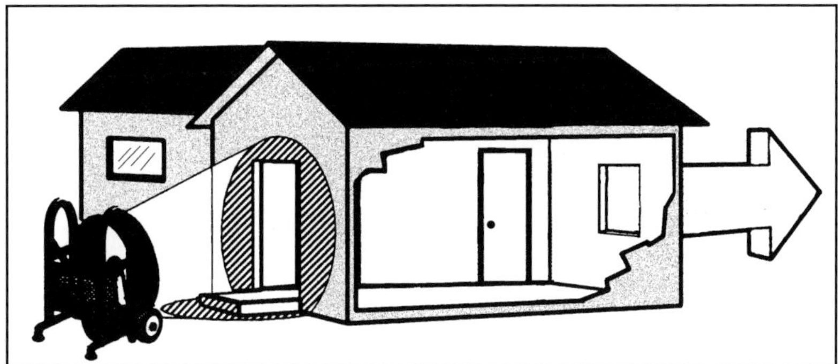

Fig. 2–6 Positive pressure ventilation uses a blower positioned outside the building to clear products of combustion after the fire is knocked down. Courtesy Tempest Technology, Inc.

Positive pressure ventilation (PPV) worked well and caught on with many fire departments as a post-knockdown ventilation technique. In recent years, specialized gas-powered blowers have been developed specifically for this purpose.

Since PPV is an important component of PPA, blower placement and other specifics will be discussed further in section II.

Water Fog Ventilation

A fog stream from a hand line directed from the interior of the structure out the top third of an opening, also called hydraulic ventilation, is another means of exhausting smoke from a structure. According to Edward P. McAniff, author of *Strategic Concepts in Fire Fighting,* a 1½-in. line will move a volume of 6,678 cfm.[2] At best, though, a fog stream will be minimally effective at clearing smoke from a structure. It is most often used to clear just a single room and almost never as part of a coordinated operation. This method benefits overhaul more than fire attack and is perhaps the least used of all methods (fig. 2–7).

Fig. 2–7 A fog stream from a hand line directed to the outside can help ventilate a single room, but is of more use for overhaul than fire attack, since it often requires using the attack line. Courtesy SLCFD

Conventional methods of ventilation leave much room for improvement, as the next chapter will discuss.

References

1. National Fire Protection Association. 2000. *NFPA 1001, Standard for Fire Fighter Professional Qualification.* Quincy, MA: National Fire Protection Association.
2. McAniff, Edward P. 1974. *Strategic Concepts in Fire Fighting.* Saddle Brook, NJ: PennWell Publishing.

3
Limitations of Conventional Methods

The previous chapter offered an overview of the generally accepted methods of ventilation and the basic operations of each. This section looks at them against the ideal established in chapter 1 for effective ventilation. Horizontal ventilation, vertical ventilation, mechanically assisted ventilation methods (negative pressure and post-knockdown PPV), and water fog all have drawbacks when it comes to how well they can be coordinated with fire attack.

Although it may not sound like it at times, this chapter is not intended to be deliberately negative or to promote one method of ventilation at the expense of another. It is a sincere attempt to take a realistic and comparative look at the norms that firefighters have accepted for generations. It will also include how these norms have been affected by changes in building materials and construction practices.

Limitations of Horizontal Ventilation

Limitations of horizontal ventilation mostly concern its relative effectiveness without being enhanced by mechanical means. For this to be most effective, the fire should be on the leeward side of the building. This tactic limits the attack entrance to the windward side of the structure and the fire to the leeward side of the structure.

Most often horizontal ventilation is accomplished by simply breaking windows in the fire area. When this is completed, the interior environment is enhanced, but not at the level that it would be if it were assisted by mechanical means.

Having crews make additional openings also takes them away from assisting with extending hoselines in support of search and rescue. Exposures near additional openings also have to be considered and protected.

Limitations of Vertical Ventilation

Vertical ventilation has been a hallmark of fire department operations for two centuries. Many a dramatic fireground photo features hard-working crews on a roof with punishing fire and smoke billowing all around them (fig. 3–1).

Fig. 3–1 Many a dramatic fireground photo features hard-working crews on a roof with punishing fire and smoke billowing all around them. Courtesy Richard Moseley

Because it happens outside in full view, vertical ventilation affords the best opportunity to capture the drama of firefighters in hand-to-hand combat with fire. Periodicals put these types of dramatic pictures on their covers because they sell magazines. Other firefighting efforts conducted outside in public view are usually in support of interior attack or are of a defensive nature, and tend to be far less visually exciting.

Without a doubt, these images of firefighters engaged in vertical ventilation communicate the danger and hazards faced every day by firefighters everywhere. After 200 years of use, vertical ventilation has had many ongoing problems that the fire service has been unable to overcome. Thus it should not be considered effective at ventilation during initial attack on residential and small commercial buildings. In most cases, it cannot measure up against the ideal of the coordinated fire attack because of factors such as the following.

The time factor

The image of firefighters responding in numbers great enough to use axes to cut ventilation holes on roofs definitely conjures up nostalgia in the fire service. But there likely were precious few successful vertical ventilation operations where the roof cut was completed prior to attack crews entering the building for rescue or an interior fire attack. The belief that roof crews can open a ventilation hole in time to be of significant benefit to fire victims or interior attack crews is basically flawed.

The example of the simulated hotel fire conducted by Southwest Research Institute was discussed in chapter 1. Although this is only one example, it shares common elements with many other similar incidents. It may also be helpful to consider the sequence of events and timing for a roof ventilation operation in this routine room and contents fire (fig. 3–2):

- The fire ignites and burns to the point where someone discovers it and calls 911.
- Companies are dispatched.
- Firefighters respond and arrive at the scene.
- Two ladders (one for roof access and one for emergency egress) are placed before crews begin working on the roof.
- A charged hoseline is taken to the roof with cutting tools.
- An opening is cut in the roof (if structural materials and building construction allow it).
- The ceiling and any interfering structures below are breached, if possible.

Realistically, can all this be accomplished using conventional roof ventilation before the victims succumb in eight minutes as determined by the study? Clearly, this would be an unrealistic expectation.

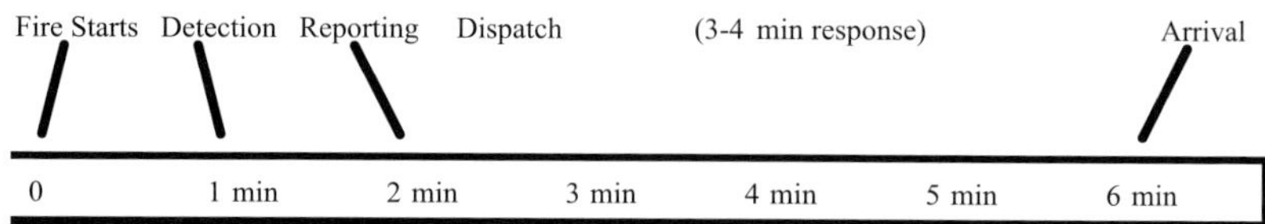

Fig. 3–2 The timeline for victim survival begins when the fire starts. Much has to happen before fire attack begins, including detecting and reporting the fire, dispatch, response, and preparing hoselines and other equipment at the scene. By the time the fire department arrives, most fires have been burning an average of five to eight minutes. In an average room and contents fire, products of combustion quickly reach lethal levels in the room of origin. (These times are estimates based on the authors' experiences.)

Firefighters in harm's way

To be most effective, vertical ventilation must be as directly above the fire as possible. For more than 200 years, this has been the objective of vertical ventilation. Unfortunately, most modern roof assemblies that fireground commanders send their firefighters on top of and into are not designed to withstand fire loads for more than a few minutes.

Certainly the roof assemblies on older buildings of ordinary construction will withstand fire involvement for a longer period of time. However, they are covered with decades' worth of roofing materials that make it difficult to open them up in time to provide any significant benefit to the initial attack crew and the victims.

Viewed objectively, vertical ventilation leaves much to chance. Crews have to pick the right location on the roof that is close enough to the fire to create a benefit. This means the fire could be burning directly below them. In such a situation, warns author Francis Brannigan, "Opening the roof accelerates the fire and weakens the structure."[1] They have to ensure the ceiling is breached. Perhaps hardest of all, they have to cut a big enough hole so the heat will not overwhelm the opening and mushroom into the attic (fig. 3–3).

Fig. 3–3 Effective vertical ventilation must consider factors such as the location of the hole above the fire, the ability of the roof to hold up with fire underneath and the weight of crews above, and how big the hole needs to be. Courtesy Ray Schelble

How much do power tools help?

One may well ask who cuts a hole with an axe anymore, anyway. It seems that today everybody uses power saws.

Power rotary or chain saws have certainly taken over. There can be no doubt that these tools are infinitely more effective and efficient than an axe in terms of reducing the effort required to cut roof openings (fig. 3–4). However, their use is limited to only a few types of building construction, such as ordinary and wood frame. For ventilating buildings of fire resistive, noncombustible, or heavy timber construction types, cutting a hole in the roof is not an option that makes tactical sense if the IC desires a coordinated fire attack.

It is true that a metal deck roof can be cut with a rotary saw. However, this is an extremely time-consuming operation and one that places firefighters on a roof system that may fail after less than five minutes of fire involvement.

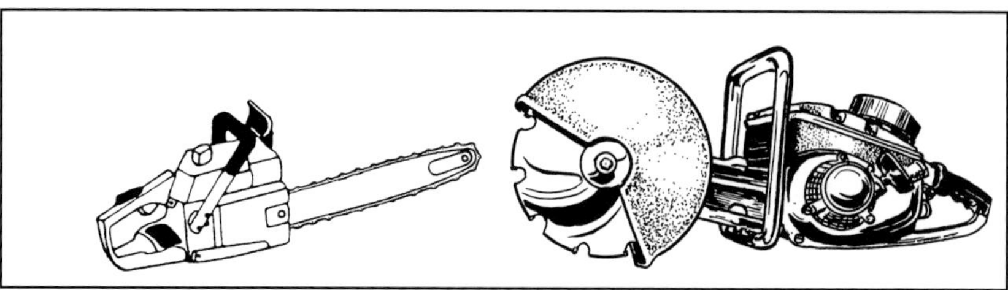

Fig. 3–4 Power tools are far more effective than an axe at cutting a vent hole. Courtesy SLCFD

Granted, compared to Benjamin Franklin's time, the basic concept of cutting a hole in a roof of a building is markedly more efficient. New technology and research on equipment have created dramatic improvements. In spite of these advances, however, it still has serious flaws when attempted as part of a coordinated attack.

Three main ventilation challenges

Construction materials and methods present three main challenges to firefighters trying to vertically ventilate a building:

- How do they deal with a fire in a building built of lightweight materials that have little mass compared to their surface area?

- How do they deal with a fire involving lightweight construction methods that, by design, will fail early when exposed to high heat and fire conditions?

- How do they deal with any structure fire considering the increased use of plastics in interior contents, which add a significant fuel load, generate excessive heat, and release extremely toxic products when exposed to heat and fire?

Of course, the answer is the early removal of substantial levels of heat and smoke and then rapid extinguishment or control of the fire. But ordering firefighters on top of a roof assembly to cut a ventilation hole is not only not the best answer, it is not even an answer at all, for two reasons.

First, to achieve anything close to a desirable effect, the vent hole must be significantly larger than most firefighters were trained to cut due to the increased heat that modern fires generate. One should consider that the rule of thumb recommending a vent hole be 10% of the size of the area involved in fire is a generation old. Without a larger vent hole, it is simply impossible to vent substantial amounts of heat and smoke. This takes time. Consequently the ventilation operation is doomed from the start to be inadequate for the interior attack.

Second, placing firefighters on a roof assembly built to fail early puts them in an extremely dangerous position. Furthermore, this might be done in a vain attempt to cut a large enough opening in a timely manner, and usually with few benefits to search and rescue and fire attack. (One should recall the definition of *coordinated attack* given previously.)

Everything has changed but ventilation

The United States Fire Administration's Fire Service Needs Workshop, in October 1999, discussed the need to better understand the effective use of ventilation. Their report stated that

> *Many of the tactics for ventilation have remained unchanged for years even though building construction and building contents have changed. These changes may result in more rapid fire growth and spread which may not be adequately addressed by existing tactics.*[2]

The most dramatic improvement in ventilation practices in a long time came with the development of a dependable, easy-to-use, gasoline-powered chain saw. As important as this innovation was, in reality it was simply the addition of a tool to perform the same job in basically the same way it had been done for more than two centuries.

Beyond that, the basic theory of vertical ventilation has not evolved. Firefighters still perform the same operations on top of and inside structures. Windows are broken and roofs are opened.

Even the older buildings that years ago would have been considered candidates for safe vertical ventilation have changed. Most of these older buildings have been repeatedly remodeled and modified with multiple roof and ceiling systems, making timely vertical ventilation of the living space virtually impossible (fig. 3–5). They are full of plastics and synthetic materials that generate high heat and toxic products levels.

Fig. 3–5 Most of the older buildings that years ago would have been considered candidates for vertical ventilation have been remodeled and modified with multiple roof and ceiling systems. Courtesy Kriss Garcia

Changes in New Construction Methods and Materials

In the past 200 years, the building industry has been evolving to meet the needs of its customers. The industry has utilized old materials in new ways and has developed less-expensive building systems that essentially keep out the elements but ignore the stress placed on them by fire. The three major building materials that have greatly influenced ventilation operations are wood, steel, and plastics.

Wood

This common building material is still used extensively, but the way it is used has changed drastically. The use of full dimensional lumber for structural elements has given way to the use of lightweight laminates and veneers that are bonded together by a variety of glues. When subjected to the heat of a fire, the plywood and waferboard sheets used for roof decking lose their ability to withstand designed loads much faster than did the solid ¾-in. wood planks in older construction.

Roof supports are often lightweight trusses with one-third the mass of full dimensional lumber and are delivered to the construction site already manufactured. These trusses depend on thin metal gusset plates to keep their compression and tension members aligned and to maintain their rigid shape (fig. 3–6). When these metal gusset plates are exposed to flame and heat, they will quickly fail. When this happens, the entire truss system becomes compromised, and the rigid truss systems will immediately fail. This failure often can involve the entire roof due to the inability of the other trusses to support the unexpected lateral loading.

Fig. 3–6 Roof supports are often lightweight trusses of one-third the mass of full dimensional lumber that are held together by thin metal gusset plates. Courtesy Kriss Garcia

Because these lightweight wood components have little mass compared to their large surface area, they burn rapidly and will consequently fail rapidly. Experience and testing show that lightweight wood decks and truss members will fail within three to five minutes of exposure to flame.

A roof assembly of lightweight trusses held together with lightweight metal gusset plates and covered with plywood decking becomes a roof collapse waiting to happen when fire is added. "There is no justification for remaining on or under wood trusses on fire," writes Brannigan. "You are playing Russian roulette. There is no 'minute rule.' There is no warning. Collapse is sudden and catastrophic."[3] Firefighters ordered onto a roof to vertically ventilate are operating on an unsafe structure that has been engineered to fail quickly, often with little if any warning.

Steel

Although steel has been used in construction for decades, the newer, lightweight steel materials, when involved in fire, can threaten firefighters' lives. Early structural collapse is a danger, and both the roof structure and roof-covering materials present problems.

Roof structure materials. One can compare the mass to surface area of solid steel structural beams with the new lightweight metal truss and bar joist assemblies. These are significantly weakened when exposed to temperatures of 1,200°F, a temperature that today's fires commonly reach. Firefighters may expect an early collapse, often without warning, of structures utilizing lightweight metal construction (fig. 3–7).

Firefighters ordered to cut a ventilation opening in a floor or roof deck made of steel will have their hands full. Certainly, given enough time and effective tools, steel can be cut. But can the operation be accomplished quickly enough to benefit the victims and the fire attack below them? The answer is no. There simply is not enough time to perform the operation and make a difference in the fire attack.

If the interior crews make a quick knockdown on the fire, the vertical ventilation operation cannot accomplish its intended purpose of helping with the attack. If interior crews do not get the fire knocked down quickly, all that heat generated by the fire continues to weaken the roof assembly while the crews work to cut a hole. It should not have to be a matter of betting the lives of firefighters on the hope that they can open a steel deck roof before the lightweight metal truss assembly fails (fig. 3–8).

Fig. 3–7 The lack of spray-on insulation in this area under the metal roof decking will hasten the failure of a roof structure that fails quickly in a fire anyway. Courtesy Kriss Garcia

Fig. 3–8 The above times are general guidelines to illustrate how long each of the five construction types can sustain fire involvement before structural failure can be expected. Notice that the common lightweight roof systems of Type II and Type V construction can fail after only minutes of fire impingement. Even when they are not directly involved in fire, the wisdom of putting crews to work on these types of roofs to cut a hole that will draw fire into the space beneath them should be seriously questioned. Roof construction features in other building types maintain strength for longer periods of fire involvement, but make it more difficult to create a hole for ventilation.

Lightweight steel building components will not support their own weight at 1,000°F and will collapse within five minutes of fire involvement.

Roof-covering materials. The roof-covering materials can also contribute to the hazards of these roofs. Flat metal roofs have several inches of fixed board foam insulation. It is impossible to sound a roof for integrity through several inches of polystyrene foam. Firefighters should remember that they only have a few minutes to complete this job (fig. 3–9). Furthermore, they may have to commit up to one-third of the on-scene resources to do so. These crews could otherwise be stretching hoselines in support of search and rescue and extinguishment.

Fig. 3–9 It is impossible to sound a roof for integrity through several inches of polystyrene foam, especially when there are only a few minutes to safely cut a hole. Courtesy Kriss Garcia

If firefighters beat the odds and are able to cut through the steel decking, another obstacle may prevent them from being successful—a membrane ceiling or suspended ceiling below the metal decking. Depending on the occupancy, these types of ceilings may be used to provide a thermal barrier from the fire below to the lightweight truss assembly and deck covering. They could also serve as a decorative or HVAC barrier. Heating and air conditioning ductwork, electrical conduits, pipes, or wiring for phones, computers, satellite dishes, and audio-video feeds may be concealed within this thermal barrier.

It is important to be very clear on this next point. It is one thing to cut a ventilation hole in a roof deck. However, being able to complete a ventilation hole through a lightweight metal roof with trusses spaced 4 ft apart and covered with several inches of insulation is something entirely different. Firefighters should consider that steel buildings having ribbed/corrugated roof decking systems are subject to collapse when any of the ribbed sections are cut, since it is these ribs that give the decking strength.

Again, firefighters must consider the time and effort required. It just cannot be done soon enough to be of any value to the people below during the initial search and rescue and fire attack (fig. 3–10). Attack lines and ventilation are of equal importance and must be accomplished together to be effective.

Fig. 3–10 The light lines on this metal roof indicate the location of the underlying roof deck supports. Cutting a ventilation hole in a metal roof deck simply cannot be done in time to be of value during fire attack. Courtesy Kriss Garcia

Plastics

The increasing use of plastics after World War II has changed the hazards firefighters face. As the use of plastics has increased, so have the toxic contaminants released in the fire and smoke, along with the levels of heat released during combustion.

To compensate for this increase in smoke and heat, it has been suggested that the roof ventilation opening be made large enough to drive a full-size automobile through. In general terms, if the BTUs in fires have increased fourfold since 1950, a 4-ft by 4-ft hole that would have been the recommended starting point in 1950 would have to be 8 ft by 8 ft to be as effective today. Using one guideline that suggests a hole should be 10% of the area involved in fire, a fire involving a 20-ft by 20-ft room would require a 40-ft^2 hole (fig. 3–11). The 4-ft by 4-ft hole is still recommended in texts today, but many, if not most, recommend making it larger, if possible.

Fig. 3–11 The openings superimposed on this roof illustrate the range of recommendations for vertical ventilation hole sizes for a room and contents fire in a 400-ft^2 room. The 4-ft by 4-ft hole on the left has been the recommended starting point for more than a century. To provide adequate ventilation for the same fire today, recommendations call for holes ranging from the size of the center hole, which at 6 ft by 6 ft is 10% of the fire area, to the largest hole on the right, which is four times the original 4-ft by 4-ft hole. Increased fire loads from modern synthetic materials are the reason.

But can a larger vent hole really keep up with the increased heat and products of combustion, especially for victims awaiting rescue? It is doubtful. The more plastics within a structure, the more fast-burning fuel will be available for combustion (fig. 3–12). It follows, then, that more toxic products of combustion and massive BTUs will be released into the interior atmosphere. This is not a survivable environment for either firefighters or civilian victims!

Fig. 3–12 Black headers like the one off this Salt Lake City fire are typical in fires involving modern, synthetic furnishings and structural materials. Courtesy John Vuyk

There are many plastics commonly used in construction and interior furnishings, including:

- Polycarbonates
- Epoxy adhesives
- Celluloids
- Acrylics
- Polyurethanes
- Polystyrene foam
- Polyethylene
- Polyesters
- Vinyls
- Nylon
- ABS pipes

These materials should make a firefighter reconsider his tactical options. This is especially true since plastic products are found in great abundance in all types of structures, including those that are fire resistive and noncombustible.

Training and Education Issues

It is the responsibility of every chief of every fire department to see that firefighters are trained to do their jobs as safely as possible. Also, every firefighter realizes that the job comes with an important responsibility to educate the public on issues vital to their safety and well-being. Yet there are problems, not only with how the fire service utilizes and trains firefighters, but also in how it educates the public. It may be helpful to consider the following:

- For most departments, vertical ventilation takes considerably more time, even using power tools, than it does to get ready to advance interior hoselines.
- Because vertical ventilation takes time, an interior attack that is coordinated with vertical ventilation will be delayed (fig. 3–13).
- An interior attack that is not coordinated with ventilation ensures that the thermal balance inside the structure will be drastically altered. Lethal temperatures and toxins will be forced throughout the structure and down to the floor level, where most viable victims are located.

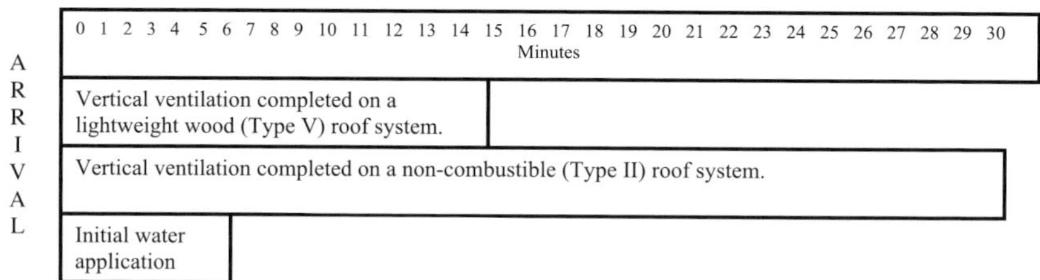

Fig. 3–13 Time is a big factor when trying to coordinate vertical ventilation with fire attack. If fire attack can begin an average of six minutes after arrival on the scene, how feasible is it for the attack crew to wait another nine minutes or more for ventilation to be accomplished?

Educating citizens

The fire service educates millions of children and adults every year that when they are trapped in a fire, they should stay low. However, without coordination with ventilation, fire attack is going to force the lethal atmosphere to the floor, where people are awaiting rescue (fig. 3–14). Firefighters must try to find them in an atmosphere in which they cannot see.

Fig. 3–14 The fire service educates millions of adults and children each year, and one of the most important lessons is to stay low if trapped in a fire. However, without coordinated ventilation, fire attack will force lethal heat and products of combustion to the floor, where people are awaiting rescue. Courtesy Gary Coon

Besides the limited chance vertical ventilation has of working in coordination with fire attack, this method also ties up firefighters with raising ladders and completing roof openings. The firefighters engaged in a vertical ventilation operation could be better utilized as additional interior crews that could be advancing attack lines in support of search and rescue operations.

Firefighters should put themselves and their families in the role of the trapped victims. Would they rather have resources deployed on the roof, or doing search and rescue?

Flawed firefighter training

Firefighter training methods are also at issue. It is common practice to use only lightweight, 7/16-in. wafer board/oriented strand board (OSB) or pallets without shingles or other roof coverings to simulate the task of cutting a vertical ventilation hole during vertical ventilation training. Most roof systems in older, larger buildings or the new lightweight metal roof systems present an entirely different problem regarding what is required to cut a hole. Training on these mock roof systems is like training for auto extrication on balsa wood cars.

Ventilation is critical to providing the life-saving functions that will allow firefighters to work in a safer environment while at the same time allowing them to make live rescues. Thus training should involve all types of roof coverings and materials on which they are likely to be working (fig. 3–15).

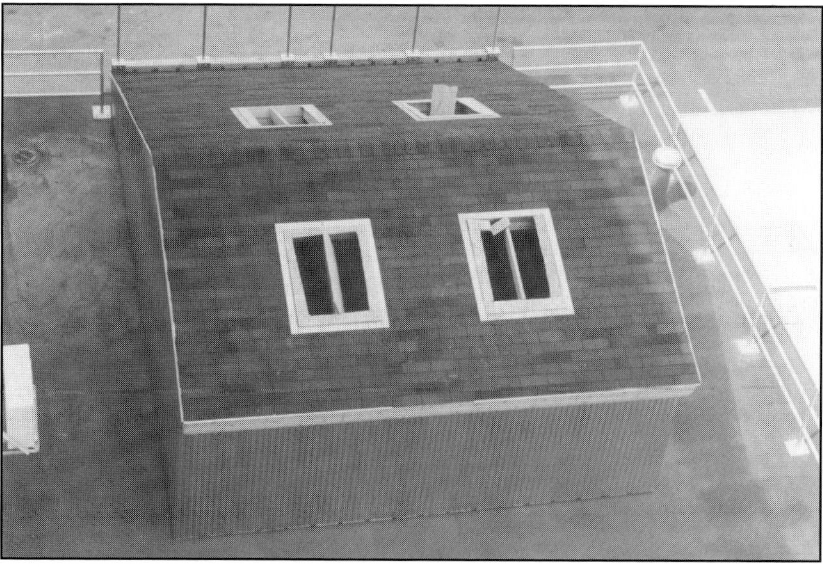

Fig. 3–15 To be of value, ventilation training on mock roof systems must include the variety of roof materials that firefighters will encounter in buildings in the areas they protect. Courtesy Ray Schelble

Once departments start to build training roofs out of the actual materials firefighters will encounter, they quickly realize how difficult it can be to provide timely vertical ventilation. To be realistic, a variety of props must be built to better simulate the buildings that firefighters will encounter (fig. 3–16). These training props might need to be 2-in. to 3-in. tongue-and-groove roof systems or plywood and lightweight trusses with several inches of tar and/or mineral surfaced roofing material. They may need a metal deck with several inches of high-density fixed board foam insulation. Departments that place this realistic element into their training realize how it alters their fireground operations.

Fig. 3–16 Training props should be designed to consider that roof systems in older, larger buildings or the new, lightweight metal roof systems present firefighters with entirely different problems regarding what is required to cut a hole. This training prop simulates a lightweight metal roof. Courtesy Kriss Garcia

Limitations of Other Ventilation Methods

For years, vertical ventilation has been the standard to remove products of combustion in the most serious situations. As can be seen, it has some obvious shortcomings, especially when it comes to the ideal of ventilation coordinated with fire attack. Although they are not as inherently dangerous as vertical ventilation, the same goes for horizontal, mechanically assisted, and water fog ventilation methods.

Horizontal ventilation

Limitations of horizontal ventilation mostly concern its effectiveness with no mechanical assistance (see next section). For maximum effectiveness, the fire should be on the leeward side of the building. Of course, the attack entrance must be on the windward side of the structure if the fire is on the leeward side.

Most often, horizontal ventilation is accomplished by simply breaking out windows in the fire area. When this is completed, the interior environment is enhanced, although not to the level of effectiveness that it would be if mechanical means were used.

Crews engaged in creating additional openings are not able to help with extending hoselines in support of search and rescue. Exposures near external openings also must be considered and protected from fire extension.

Mechanically assisted ventilation

Powered fans and blowers can help clear a structure faster than leaving it to natural currents or fire dynamics and air flow. However, they still cannot meet the ideal for coordinated ventilation.

Negative pressure ventilation. Although adding fans provides an improvement over unassisted ventilation, this method still cannot make an effective coordinated attack. First, the concept of "sucking" air out of a building can only be successful if the building remains mostly closed up. Negative pressure can only be drawn if a near vacuum has been achieved and replacement air coming in is less than the air being exhausted. After windows have been broken or doors opened, smoke ejectors have a difficult time overcoming the atmospheric pressure outside the building to produce adequate ventilation. Second, since most smoke ejectors are electric powered, they require a remote power source such as a generator—another equipment factor to delay ventilation (fig. 3–17).

Like vertical ventilation, smoke ejectors cannot meet the goal of coordinating ventilation with fire attack. Because of how they work, they are most often set up after the fire is knocked down. They have advantages for overhaul but are not really effective for rescue or fire attack.

Fig. 3–17 Smoke ejectors will have a difficult time clearing an area of heat and products of combustion after windows have been broken and doors opened. Also, since they are electric powered, they require a power source. Courtesy Kriss Garcia

One additional drawback of smoke ejectors is that they are usually positioned in the doorway used to enter and exit the building. For maximum effectiveness, they have to be set up where most of the heat and smoke is, which is near the ceiling. This requires equipment such as hooks, rope, and ladders, and also requires additional time to put everything in place.

Post-knockdown PPV. PPV can efficiently remove products of combustion from a structure. It is quick, and it can be implemented by simply positioning a blower in a doorway. However, post-knockdown PPV cannot be considered part of a coordinated attack, since it is not accomplished at the same time fire crews are advancing on the fire (fig. 3–18). Instead, crews operate in a flammable and toxic environment as they attempt to complete rescue and fire control and then begin ventilating afterward.

Typically, fire departments that are moving toward PPV purchase several blowers and place them on their apparatus. The blowers are put into operation only after crews are sure the fire is out.

Fig. 3–18 Post-knockdown PPV cannot be considered part of a coordinated attack since it is not started until after the fire is knocked down. Courtesy Richard Moseley

As recently as 2002, fire journals were still publishing articles encouraging firefighters to use PPV, but only to remove "cold smoke" after the fire is out. This practice has three major flaws:

- Steam-laden soot impregnates all of the property owner's belongings. If after a fire there are blackened walls throughout the structure, and not just in the immediate area of fire involvement, one can be assured that a coordinated attack was not made (fig. 3–19).

- When firefighters crawl through a structure to a point between the fire and a ventilation opening, hot and toxic products of combustion may be forced into where they are working when the blower is started. This situation has the potential to trap firefighters in an atmosphere in which they cannot survive. It would be a similar situation to placing firefighters in the attic and then opening up the roof and ceiling for vertical ventilation. Firefighters would never think of doing such a thing, yet with lack of training or miscues on the fireground, it could happen.

- When victims are located between the fire and the ventilation opening, they face dangers similar to the firefighters in the previous example. But these victims will still be better off with PPV than they would be with other ventilation methods. Even when used post-knockdown, the ability of PPV to efficiently remove heat and products of combustion will quickly replace the toxic atmosphere with one that is cleaner and more survivable. The true hazard in this situation is the disturbance to the thermal balance initiated by the fire attack without ventilation, not the post-knockdown PPV. Firefighters should remember that the same soot that stains the walls and belongings is being forced into the areas where victims are located. They have no option other than to breathe the lethal combination of heat and toxic products of combustion.

Fig. 3–19 Smoke and soot stains on the walls and belongings are inevitable in the area of the fire. Ventilating late in a fire can spread this type of damage throughout a building. Courtesy TCVFD

Water fog

Using a fog stream to provide ventilation for fire attack is not an option that makes much sense for coordinated fire attack. It ties up firefighters the whole time it is in operation. Most often, this method requires the very same hoseline for ventilation that was used to knock down the fire. Since interior crews must use the attack line to force products of combustion outside, that line is not helping with overhaul.

Additional Factors to Consider

Beyond the points already presented, other factors relating to conventional ventilation methods warrant consideration.

Who ventilates?

In larger departments, ladder truck companies are traditionally assigned to do ventilation. As discussed earlier, vertical ventilation is commonly the method of choice.

The problem they encounter is two-fold. First, most departments have more engines than truck companies, sometimes by a great margin. Because of this, trucks are often second or third in on an incident and may not arrive until after the fire attack has begun (fig. 3–20).

Fig. 3–20 Because most departments generally have fewer ladder trucks than engines, they may not even arrive at a scene until after fire attack has begun. Courtesy Richard Moseley

Second, generally speaking, vertical ventilation takes a minimum of approximately 10 to 20 minutes from the beginning of the operation until the point where smoke and heat can vent. On the majority of fires (residential and small commercial), it generally takes only a few minutes to accomplish an aggressive interior fire attack and apply water to the fire. Firefighters tend to be action-oriented people anyway, and where there is urgent work to be done, engine crews simply will not wait for a truck crew to open the roof. The interior attack begins as soon as attack lines are pulled and water supply is secured. Crews battle against the fire as well as the punishing smoke, heat, and darkness to find victims and the seat of the fire.

In smaller departments, which make up the majority of the departments in the United States, second-arriving companies generally perform ventilation duties. They face problems that are much the same as those of larger departments.

Whether a department has 5,000 firefighters or five, ventilation cannot be coordinated with fire attack in support of search and rescue if one operation takes five minutes and the other one takes 10 or 20 minutes. This is especially true if ventilation is assigned to later-arriving companies.

Mechanically assisted ventilation has similar problems. By the time truck companies or later-arriving companies deploy the blowers, first-in crews are already inside applying water. This displaces the deadly atmosphere to areas where victims are trying to survive. Plus, the firefighters' job of advancing on the fire is less effective and much more hazardous due to the heat, smoke, and decreased visibility.

Tactical considerations

We have established that ventilation coordinated with fire attack can provide a wide range of benefits to firefighters, victims, and property owners. So why do the available lists of tactical objectives fail to reflect this importance?

Many acronyms have been developed over the years to help firefighters determine a priority sequence of tactical objectives. Ventilation has historically been plugged in somewhere in the later part of the sequence near salvage and overhaul. The most popular acronym in use today, which also is used by the National Fire Academy, is RECEO-VS. This stands for rescue, exposure, confine, extinguish, overhaul, and ventilation and salvage. Developers of this acronym intended to illustrate that ventilation and salvage could be started at any point in the sequence. However, the order of the elements in the acronym tends to give firefighters the mistaken impression that ventilation and salvage are either the last objectives or are not as important as the earlier ones. This idea is wrong, as a coordinated attack requires that hoselines and ventilation be given equal importance and be accomplished simultaneously. If ventilation must be operating to provide the benefits of a coordinated attack, why has the fire service consistently given ventilation the status of a lesser priority?

The answer seems clear. Historically, ventilation follows fire attack. Since this places firefighters and victims in a toxic atmosphere, it strongly suggests the fire service in general has not emphasized conducting a ventilation operation during the time it will provide the greatest benefit.

To address this shortcoming, the authors suggest that RECEO-VS is a simple and still appropriate acronym to define tactical priorities, but with a modification (fig. 3–21).

When ventilation is given an appropriate status in overall fire operations, it starts at the beginning of fire attack and continues until the end of the operation.

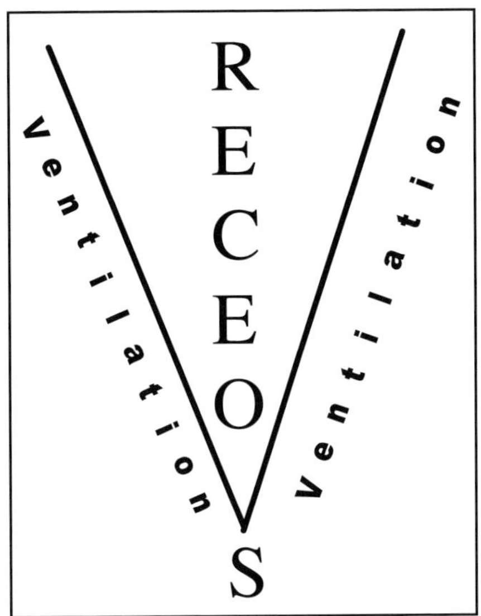

Fig. 3–21 The National Fire Academy acronym RECEO-VS for fireground tactical priorities includes: rescue, exposure, confine, extinguish, overhaul, and ventilation and salvage. The authors suggest the modification illustrated here to emphasize the expanded role ventilation should play in the operation.

Increased BTU evolvement

As the use of synthetic materials and plastics continues to increase at an exponential rate, so do its impacts on effective ventilation. Because of the extremely flammable nature of these materials, they should be considered essentially as solid gasoline when exposed to fire. Furnishings, adhesives used in construction, and window coverings, decorating, and construction materials all may have components made from hydrocarbon-based materials. They give off massive amounts of suspended toxic materials when decomposing in a fire. In addition, they also increase fire spread. According to author Stephen Ames, "Burning droplets can spread downwards and substantial acceleration of fire growth is often produced by the formation of pools of burning liquid beneath the item involved. This is a particular problem with domestic upholstery where thermoplastic deformation of polyolefin fabrics and the decomposition of polyurethane foams produce flammable liquids which significantly accelerate the fire growth rate."[4] Plastics can also generate up to 20 times the heat as previous materials.

As discussed earlier, a 4-ft by 4-ft hole was once considered an adequate starting point to ventilate the BTUs and products of combustion created by furnishings and items made of natural materials. This size of hole is no longer adequate. In "Thermal Environments of Structural Fire Fighting," J. Randall Lawson reports, "Fire growth characteristics have changed significantly over the last 50 years in North America. These changes can be attributed to: 1) increases in compartment fire loads, 2) increases in rates of heat release for these fire loads, and 3) differences in building construction."[5]

Then, as now, it was recognized that if fire and smoke vented forcibly from a vent hole, the opening was not large enough. Not providing adequate area in a vent hole to allow products of combustion to vent freely creates conditions that encourage fire to extend throughout a previously uninvolved attic. It also means the efforts of the ventilation crews have contributed little to the favorable outcome of the incident.

Anyone who has spent tense moments operating on the roof over a structure fire understands the problem. With the time it takes to open a hole large enough to allow the majority of BTUs to escape in time to support interior operations, it may simply not be a practical response.

Difficulty with roof coverings

The elements of a roof that make it impervious to weather can present major obstacles to any roof ventilation operation. Because of the way they are built, buildings of ordinary construction are generally considered one of the safer types for roof operations. However, these buildings are decades old. Years and years of reroofing and remodeling can add layers of roof and ceiling materials that make it difficult to cut through in a timely fashion.

In lightweight metal construction, the steel decking and insulation can be very difficult to cut completely through. When there is a heavy fire load, it is an impossible task to complete within the five minutes or less before the failure of structural members could lead to building collapse. When there is not a substantial fire load, the IC must weigh the potential benefits from the assignment against the risks.

Ventilation through roof coverings will be discussed in more detail in the later section on building construction types.

Cannot push down ceiling

Perhaps a crew has overcome all the problems with the roof covering and construction. After working hard to ladder the roof, extend a hoseline, and cut a hole big enough to vent the massive amount of heat below, they wonder, "Where is the smoke?"

Many times crews cannot complete the operation because they cannot push down the ceiling. For vertical ventilation to be effective on a fire in a living space, the ceiling must be opened to the same dimensions as the hole that was cut in the roof (fig. 3–22). If firefighters make a 4-ft by 4-ft roof opening but are only able to make a 2-ft by 2-ft opening in the ceiling, the effectiveness of the ventilation has been significantly reduced. Much emphasis is put on opening the roof, but the ceiling opening is also critical to success.

Fig. 3–22 For vertical ventilation to be effective on a fire in a living space, the ceiling must be opened to the same dimensions as the hole that was cut in the roof. Courtesy SLCFD

There are many reasons why crews may not be able to complete this final phase of the operation (fig. 3–23). Sometimes the construction of the ceiling simply makes punching a hole difficult. Wood and metal lath in the ceiling can interfere with pushing down a large enough area. Or the ceiling could be made of decorative solid wood material that is next to impossible to push down from above.

The attic space may have planking installed on top of the ceiling joists for storage of items such as family heirlooms or business records (fig. 3–24). Attic spaces are also full of utility construction for wires, piping, cable, and HVAC plenums that can be serious impediments to completing a vent hole.

Fig. 3–23 After completion, the roof remodeling on this building will make it extremely difficult to complete a vent hole through the roof to the interior. Courtesy Kriss Garcia

Fig. 3–24 Multiple ceilings can turn a vertical ventilation operation from difficult to impossible. Courtesy Kriss Garcia

The simplest reason of all for not being able to complete a vent hole may be that crews cannot reach down far enough to reach the ceiling. If tools cannot reach far enough to open the ceiling, the roof opening is of little or no use for ventilation.

Two things happen when a vent hole cannot be completed:

- The roof crews work hard for quite some time on a dangerous operation that most likely still would not be completed in time to fulfill tactical objectives for attack and victim survival.
- Now all their efforts will not help substantially with overhaul, either.

A ventilation crew working directly above the fire, as is the goal, may only get a few attempts to push down the ceiling, as fire will start to vent from the hole. Sometimes after punching through a ceiling, fire and high heat do not exit immediately but progressively work their way to the hole. In this case the ventilation crews have been successful only in assisting the fire in its search for fuel and negative pressure as it extends through the structure to reach the hole.

Compartmentalized construction

Author Kriss Garcia tells the following true story. He relates that "working as a member of a company assigned to vertical ventilation on a house fire, I was backing up a firefighter who was cutting the lightweight roof decking material. After the initial hole was cut, I started punching down the ceiling with a pike pole. As I pulled the pike pole up between punches, I noticed something on the end of the pole. I had stabbed a box of Cheerios. On my next attempt, I brought up a box of oatmeal."

This illustrates another complication with vertical ventilation. Sometimes firefighters can make a cut through the roof covering before the roof structure has been compromised and can reach the ceiling to punch a hole. However, they still may be into an area that will not allow them to ventilate the fire area (fig. 3–25).

As another example, at 3:00 a.m., firefighters on a roof are making their way to a position as close to above the fire as they believe is safe. They start cutting the vent hole. They have been trained to get as directly over the top of the fire as possible, and common sense says this is as close as they dare get. If their estimate of the fire's location is off even a few feet, and they end up on the other side of a partition, they will get some heat and smoke out. However, it will not be enough or in time to even come close to be coordinated with attack. Or they may end up cutting a vent hole over cabinets, soffits, closets, or other compartments that do not open into an area that will vent the fire. A review of several generic house plans showed that approximately 10% to 20% of the ceiling space had installed barriers that would impede successful vertical ventilation.

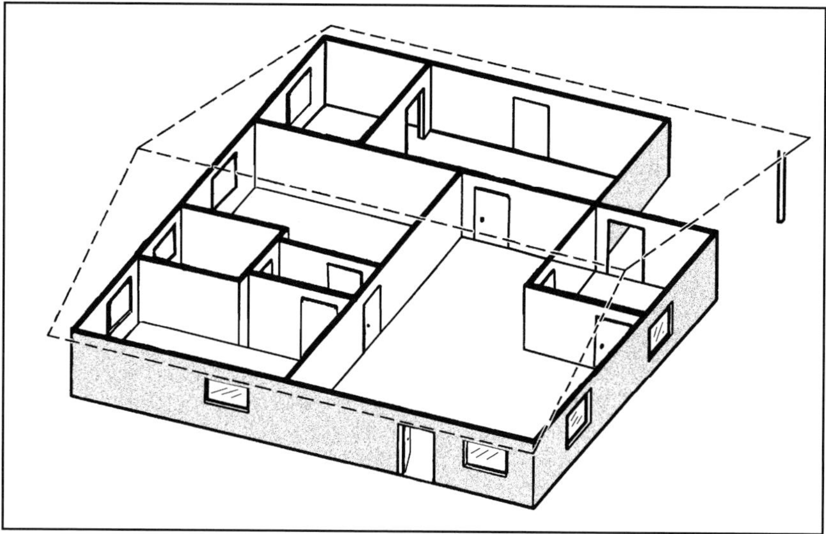

Fig. 3–25 A building that is divided into rooms is compartmentalized. Cutting a vent hole in a closed space or compartment that is separated from the fire area will not provide effective vertical ventilation. Courtesy Tempest Technologies, Inc.

Power tools may not work

Another factor that victims and interior crews depend on for their safety and survival is that power tools will function properly. Modern firefighters truly depend on their power tools, and the inability to use them during roof operations will seriously delay any removal of heat and smoke. Any firefighter can recall incidents where the tools they were using did not function properly. Perhaps crews are making an interior attack when challenges with power tools on the roof cause a delay in an operation that already takes a considerable amount of time. In these cases life-threatening consequences are more assured for those awaiting rescue.

Firefighters may fall into the fire

Ventilation crews must make their way across the roof to a point as directly above the fire as possible. In the majority of single occupancy or small multiple dwelling fires, firefighters are operating on roof systems that are likely to fail within only a few minutes. Crews that open a vent as directly above the fire as possible and then are not able to make a large enough opening provide no benefit for themselves and other crews fighting the fire. They have actually made it easier for fire to move into previously uninvolved areas under the roof and above the ceiling. More of the already compromised roof they are standing on will be negatively impacted. The materials on the majority of roofs on new commercial buildings are not designed to withstand fire involvement for more than a few minutes (figs. 3–26 and 3–27).

Many firefighters have seen the video of Phoenix firefighters on Ladder 27 falling through a tile roof on a house fire ventilation operation. Many have also seen footage of the California firefighter falling through the flat roof of an apartment building. These lightweight roof systems have been tested by private and governmental agencies and have been shown to be prone to early collapse.

Deputy Chief Vincent Dunn of the New York City Fire Department writes, "Firefighters should understand that, when they walk upon the roof of a burning building, they risk the possibility of plunging through a fire-weakened structure into a dark, smoky, flaming, cage-like enclosure of four walls from which they will not escape alive."[6]

Fig. 3–26 Figures 3–26 and 3–27 show the same roof from the outside and the inside. In this figure, the roof is still intact on the outside. Courtesy Richard Moseley

Fig. 3–27 But on the inside, fire has destroyed all the supporting structure underneath it. It would not have been able to support the weight of a firefighter long before it got to this point. Courtesy Richard Moseley

Sprinkler system activation

Although sprinklers provide a great benefit to fire control, activation of an automatic sprinkler system causes conflicts with vertical ventilation. The activated sprinklers cool the atmosphere inside the building, which decreases the tendency of products of combustion to rise. A ventilation hole in a roof will not be of much help if the gases inside will not rise.

In one example, crews from a Florida department worked for more than one hour attempting to make a 4-ft by 4-ft vertical ventilation hole in a large, noncombustible structure where the sprinklers had activated. At one point a saw blade sharpener was called to the scene to sharpen the blades used by the vertical ventilation crews. When the hole was finished, the products of combustion inside had been cooled by the sprinklers to the point where they would not rise through the ventilation hole (fig. 3–28).

Fig. 3–28 Activated sprinklers cool smoke and gases inside a building, which decreases their tendency to rise. Vertical ventilation will not work on an atmosphere that cannot rise. Courtesy Kriss Garcia

Ventilation by Building Construction Type

Type I: Fire resistive

Type I buildings are also called fire-resistive buildings (fig. 3–29). They are generally high-rises and could be businesses, apartments, or other high life hazard type structures. They are primarily built of steel structural members that are covered by reinforced concrete (fig. 3–30). The roof assemblies are formidable, made of reinforced concrete decks. The very nature of this type of construction does not allow for breaching the roof to provide additional vertical ventilation.

Characteristics of Type I Fire Resistive Construction

Structural Elements	Steel with a fire-protective covering; reinforced concrete.
Floors	May be poured-in-place concrete, prefabricated slabs, or other fire resistive material.
Walls	• Exterior walls may be precast lightweight concrete, aluminum, glass, or other material. • May be bearing walls or may not be. • Walls have varying fire resistance. • Stairwells enclosed in fire resistive materials.
Roof	May be similar to material used in floors.

Fig. 3–29 The above identifies building components for various structural elements in fire resistive consturction.

Fig. 3–30 Type I buildings are generally high-rise structures built primarily of steel structural members that are covered with reinforced concrete. Courtesy Kriss Garcia

The best way to provide ventilation in these types of structures is to utilize and augment the building's fire protection systems, including the smoke removal systems that are designed into them. Many have automatic smoke control systems in place to assist in removing products of combustion. It may be unrealistic to believe anything more can be done beyond supplementing and supporting these designed ventilation systems to meet immediate tactical objectives.

Type II: Noncombustible

Type II construction is also called noncombustible construction (fig. 3–31). This type of construction is characteristic of the majority of lightweight metal roof systems being built over both small and large commercial and some residential buildings (fig. 3–32). The primary hazard of roof operations on these types of buildings is early collapse of the unprotected steel roof assembly. Tests have shown that following fire involvement, integral elements of these systems will fail within five minutes. Firefighters and ICs must realize that sending crews on top of or inside these types of structures should have a critical, absolutely necessary life-saving tactical objective as the goal. It is imperative that they first perform a thorough incident risk analysis of the building and that the potential life hazard be weighed against the construction type, burn time, and the extent of fire involvement.

Characteristics of Type II Noncombustible Construction

Structural Elements	Exposed steel without fire-protective covering, with bolted, riveted, or welded connections.
Floors	May have lightweight steel supports.
Walls	• May be masonry, steel, aluminum, glass, or other material. • Wall construction materials may be hidden by wall coverings.
Roof	May have lightweight steel supports.

Fig. 3–31 The above identifies building components for various structural elements in noncombustible construction.

Fig. 3–32 The majority of lightweight metal roofing systems are on buildings classified as Type II construction, including small and large commercial and some residential buildings. Courtesy Kriss Garcia

Even if the risk analysis warrants rooftop operations, firefighters must recognize that making an appreciable hole in this roof system will take a considerable amount of personnel and time. These roof systems usually have high-density insulation decks that are several inches thick. As discussed earlier, the ability to sound the roof to determine weakness is greatly hampered by the insulation deck, as is the ability of the crews to get through to the metal roof deck to start making a hole.

Cutting a metal deck and then removing material that is fastened with screws into the truss systems is a very time-consuming process. Perhaps the most dangerous situation comes if the operation is successful. If fire starts venting from the hole, the effects of heat on the roof structure are accelerated. The time to complete the objective and safely exit the roof before a catastrophic collapse will be greatly reduced (fig. 3–33).

Fig. 3–33 A lightweight metal truss has little mass and will not withstand high levels of heat for long without failing. Courtesy Kriss Garcia

Depending on the type of metal roof system, the deck itself is often an integral part of the roof structure that, when cut, will compromise the roof's ability to support loading. Since structural supports and lightweight metal trusses are often spaced 4 ft apart on these roofs, cutting the deck can compromise the entire area on which firefighters are operating (fig. 3–34).

In other words, the deck is a structural member, and cutting it may have the same effect as Wile E. Coyote cutting off the branch on which he is standing.

Fig. 3–34 Since lightweight metal trusses are often spaced 4 ft apart, cutting through a roof deck can compromise the entire area on which firefighters are operating. Courtesy Ray Schelble

Type III: Ordinary

Buildings of ordinary construction are most often a minimum of several decades old and pose a particular hazard for firefighters (fig. 3–35). Brannigan states, "More firefighters have been killed in fires in ordinary construction buildings than in any other type. Fire officers should become more analytical of the hazards of this type of structure and rely less on experience which may not be adequate."[7]

Characteristics of Type III Ordinary Construction	
Structural Elements	Masonry and wood
Floors	Usually wood supports with wood decking.
Walls	Masonry bearing walls, reinforced with steel starting in the 1960s. Interior walls have heavy lumber supports
Roof	**Conventional:** Usually heavy wood supports with wood decking. **Reinforced:** Lightweight wood supports with wood decking.

Fig. 3–35 The above identifies building components for various structural elements in ordinary construction.

The authors have broken ordinary construction down into two categories: conventional and reinforced. The fact that these buildings have been around for at least several decades almost ensures they have had several roof coverings. Many have additional roof and ceiling structures due to remodeling, repair, and maintenance (fig. 3–36).

Fig. 3–36 The fact that buildings of Type III or ordinary construction have been around for at least several decades virtually ensures that they have had several roof coverings. Many have additional roof structures due to repair and maintenance. Courtesy Kriss Garcia

Construction characteristics. Buildings of conventional ordinary construction date from the late 1800s up to approximately the 1950s. This type of construction is characterized by unreinforced masonry walls. They also have floor joists with a "fire cut," so if the joists fail, they can pull out of bearing walls. This feature is designed to make a catastrophic wall collapse less likely (figs. 3–37 and 3–38).

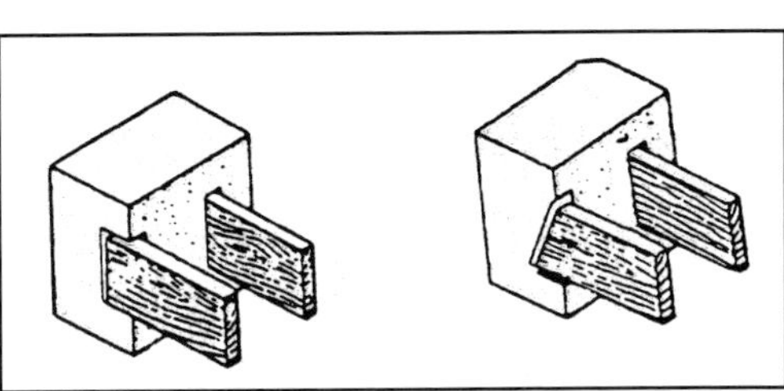

Fig. 3–37 A "fire cut" is typical in the floor joist of conventional Type III buildings (right). This allows the joist to pull out of the wall if the joist fails, which makes catastrophic wall collapse less likely. Courtesy SLCFD

Fig. 3–38 The masonry wall in this remodeled Type III building clearly shows the slots for the fire cut on the since-removed floor joists. Courtesy Kriss Garcia

These buildings have massive unreinforced masonry walls with full dimension wood floor and roof systems. The roof systems are substantial, with full dimensional lumber supporting tongue-and-groove roof decks. The burn time before failure is approximately 20 minutes, assuming no deterioration such as dry rot exists and no alterations have been made that could weaken the structural components. The primary issue that negatively affects vertical ventilation of these buildings is related to remodeling and repairs done over the years.

Reinforced ordinary construction has steel reinforcement in the masonry walls and lighter weight, nominal dimension wooden joists. Walls may be tilt-up, poured concrete, or cinderblock. These are generally smaller commercial buildings, since the minimal mass of the lumber used must be able to adequately support load-bearing supports across the span.

The roof decks of reinforced ordinary buildings are generally supported by nominal thickness lumber. The decking material is generally wood panels or plywood laid over and fastened with nails to the joists. Multiple tests made by various agencies have shown that the burn time of this type of construction averages seven minutes before structural members fail. In reinforced ordinary constructed buildings, there is no possible way for a company to complete a vertical ventilation operation on the roof before their survival is jeopardized.

Ventilation challenges. Being subjected to decades of weather and aging, most roofs in both types of ordinary construction have had several episodes of major structural modifications, repair, and/or maintenance. Many have had multiple roof coverings applied over the years and may have additional "rain roofs" built over existing roofs. Of course, remodeling of these buildings is not limited to the roofs. Additional ceilings are often built below the original. Both conditions dramatically impede vertical ventilation.

Multiple roof coverings make cutting through several inches of full dimension tongue-and-groove decking and several inches of tar, gravel, and other weather-resistant membranes next to impossible. In addition, these materials will quickly damage the blades of power tools, and hand tools are virtually useless in making a substantial ventilation hole. All this makes performing a timely coordinated fire attack nearly impossible. (As stated previously, ventilation is not coordinated unless it actually removes the products of combustion in coordination with the attack.)

Rain roofs built over existing roofs add an additional layer and create horizontal open space. Once these roofs are opened, false ceilings, cocklofts, or attic spaces can prevent crews from opening the ventilation hole to the interior space to vent the fire area. Sometimes the vertical ventilation hole cannot be located directly over the fire and cannot be made large enough to allow all the BTUs generated by the fire to escape. Unfortunately in these cases, the firefighters' efforts will in effect have opened up a path for the fire to extend into these previously uninvolved open horizontal areas. Bringing fire and smoke into these combustible-loaded spaces encourages fire extension throughout the building.

Another consideration for firefighters engaged in vertical ventilation in buildings of conventional ordinary construction is the collapse of parapet walls that extend above the roof. The mortar on these old buildings quickly deteriorates during fire conditions. Any deformity in the wall may cause the parapet to fall, hitting firefighters engaged in roof operations or causing the roof to collapse due to the impact loading of the parapet materials.

All these conditions pose challenges that make vertical ventilation in these structures an undependable option when a coordinated fire attack is needed.

Type IV: Heavy timber or mill construction

From a vertical ventilation standpoint, heavy timber construction is very similar to conventional ordinary construction (fig. 3–39). One exception is that the decking materials are much thicker and the buildings are older, dating back as far as the early 1800s (fig. 3–40). These buildings have roof decks made of tongue-and-groove planking that is 2 in. to 4 in. thick and has been subjected to several more decades of repair and remodeling. This makes providing a timely vertical ventilation hole even more difficult.

Characteristics of Type IV Heavy Timber Construction	
Structural Elements	Heavy lumber and masonry.
Floors	Wood usually at least 3 inches thick.
Walls	Solid wood or laminated wood.
Roof	Heavy wood supports with wood decking.

Fig. 3–39 Heavy timber type construction, also known as mill type construction.

Fig. 3–40 From a vertical ventilation standpoint, Type IV heavy timber construction is very similar to Type III ordinary construction except that the decking materials are much thicker and the buildings are older. Courtesy Kriss Garcia

These buildings are more substantial than ordinary construction in terms of mass and will withstand fire involvement for a greater period of time. Like conventional ordinary construction, the exterior walls are massive and made of unreinforced masonry. Much of the mortar of this period was made with a sand-lime mixture because there was a shortage of cement products available to provide adhesion and strength. Wall stability should be of concern to firefighters and incident commanders. Catastrophic wall collapses are always a possibility as these buildings degrade throughout a fire event (fig. 3–41).

Fig. 3–41 Because they are substantially built, Type IV buildings will withstand fire involvement for a greater length of time than Type III, but catastrophic collapse of the unreinforced masonry walls should be of concern to firefighters. Courtesy Kriss Garcia

Type V: Wood frame

Buildings of wood frame construction should give concern to firefighters (fig. 3–42). These roofs have been popular in residential and small commercial buildings since the 1940s. They are made of either lightweight wood trusses held together with metal gusset plates or stick-framed with nominal dimension lumber (fig. 3–43).

Characteristics of Type V Wood Frame Construction	
Structural Elements	All structural supports made from wood.
Floors	Wood.
Walls	Wood covered with various materials, including gypsum board (drywall) and plaster.
Roof	Wood support structure and decking.

Fig. 3–42 Typical wood frame construction where all structural components are of wood.

Many roof systems of this type are on residential buildings and have some slope to the roof. Deputy Vincent Dunn of the New York City Fire Department states, "When firefighters are operating at a private home with a sloping roof, it is more effective to remove smoke from the structure by venting a top-floor window than by cutting a vent opening."[8]

Fig. 3–43 Roofs on Type V wood frame construction are typically supported by lightweight wood trusses held together with metal gusset plates, as in this photo, or stick-framed with nominal dimension lumber. Courtesy Kriss Garcia

The major compromising factor is the decking that is installed on top of the structural roof member, which is typically either plywood or OSB. Both behave similarly under fire conditions. When heated, the decking degrades to the point that it cannot hold the dynamic load of firefighters. This decking is usually ⅜-in. to ⅝-in. thick and is unable to support a firefighter after three to five minutes of direct fire involvement (fig. 3–44).

The structural supports will withstand fire loading for up to 10 minutes. However, it is the decking material that firefighters operate on and count on for support that rapidly gives way. This not only jeopardizes their safety, but also the safety of others whose lives may be depending on the ventilation being successfully accomplished.

Fig. 3–44 The decking installed on top of the structural roof members, typically either plywood or OSB, degrades to the point where it cannot hold the dynamic loads of firefighters within about three to five minutes of fire exposure. Courtesy Kriss Garcia

The wood truss members have the same limiting factor as the decking, with one big complication. They are part of an interdependent truss system that is held together with metal gussets. These gussets tend to loosen and fail after only five minutes of fire involvement. As with all truss systems, if any one member fails, tension and compression loads, as well as lateral loads, are shifted to other trusses. These trusses are not designed to withstand the undesigned shifting of the loading, and the possibility of a catastrophic collapse is real.

This type of roof system is the quickest to vertically ventilate. It will also fail the fastest. If it takes proficient crews seven to 10 minutes on average to complete vertical ventilation on wood frame structures, it is still several minutes too long. In addition to these hazards, by the time crews complete vertical ventilation, interior crews have already entered, with the potential for forcing the toxic atmosphere down to where victims are waiting for rescue. Another consideration is whether the attack is extending the fire into a previously uninvolved area, such as the attic.

Aspects of Indirect Attack

Indirect attack is a tactic that does not involve ventilation but still deserves mention here. It was developed to increase the efficiency of fire attack in certain situations.

The attack is made from a position outside the area involved in fire. Areas involved in fire are kept closed. Water is directed up into the hottest part of the space, thereby blacking out the fire as it is converted to steam and expands.

This method, while effective for some types of fire situations, does not address the need to make timely and viable rescues. It also exposes firefighters who may be inside to punishing heat and smoke.

The effect is essentially the same as what happens when interior crews attack the fire before the interior is ventilated. The only difference is whether firefighters and victims are inside or outside the confined space. Anyone inside is subjected to the damaging effects of loading the heated atmosphere with moisture, which makes the environment far more lethal for victims.

An Alternative

By its nature fighting fires is demanding, dangerous work. Even if every conceivable safety innovation were implemented and followed to the letter, it is impossible to envision a time when firefighters could do their job without some getting injured or killed.

With the United States having the highest death and injury rate from fire of any industrialized nation, this nation's fire service must continually ask if there is a better way. Firefighters have adopted many changes over the years that have increased safety for themselves and the victims of fire, even though those changes were not always accepted as fast as some would have liked.

To be fair, the discussion of ventilation as it relates to fire attack has been ongoing and has led to improvements. However, this discussion has been confined largely to refining methods that have been around for years—methods that will not, with any sort of consistency across the fire service, approach the maximum benefit that can be expected of ventilation. For all the discussion and knowledge, it still does not take all or even a few of the conditions and situations discussed in this chapter to make any method of ventilation too late or too little to benefit fire attack and victims. It only takes one.

But there is an alternative to consider.

PPA is a fireground tactic that has the potential to address these concerns over a wide range of situations better than conventional or post-knockdown ventilation methods, as will be seen in the next section.

References

1. Brannigan, Francis L. 1999. The unexpected killer. *Fire Engineering*. 152 (1) (January).
2. National Institute of Standards and Technology. 2000. Fire research needs workshop proceedings. Gaithersburg, MD: National Institute of Standards and Technology.
3. Brannigan. 1999.
4. Ibid.
5. Lawson, J. Randall. 1997. Thermal environments of structural fire fighting. Firefighter thermal exposure workshop: protective clothing, tactics, and fire service PPE training procedures. June 25–26, 1996. Gaithersburg, MD: National Institute of Standards and Technology.
6. Dunn, Vincent. 1988. *Collapse of Burning Buildings*. Saddlebrook, NJ: PennWell Publishing.
7. Brannigan, Francis L. 1992. *Building Construction for the Fire Service*. Quincy, MA: National Fire Protection Association.
8. Dunn. 1988.

II
Positive Pressure Attack

Rescue and fire attack are the cornerstones of a firefighter's job. Until crews announce an all-clear and put the fire out, pretty much everything that happens on the fireground is in support of these efforts.

As discussed in the previous section, clearing products of combustion from the interior of a building as early as possible deserves the highest priority. The safest and most effective way for firefighters to achieve the ideal of ventilation coordinated with fire attack is by using PPA. PPA is based on the principles of PPV, which has been briefly discussed in earlier chapters. Section II will provide more detailed instruction in how to use PPV, and then will make the jump to using PPA. Chapter 6 clarifies what precautions must be observed for safety and effectiveness.

Firefighters should not attempt hands-on training with PPV or PPA until they read and understand chapters 4, 5, and 6, and until all involved personnel have received training in its use.

4

Understanding Positive Pressure Ventilation

For the purposes of this text, positive pressure ventilation is the application of positive pressure by using mechanical means *after the fire has been knocked down*. This is the way positive pressure is used by most fire departments. Positive pressure attack, on the other hand, requires positive pressure be implemented before attacking the fire, and it will be discussed at length in the next chapter.

PPV has been in use to some extent for almost 50 years. It has proven extremely effective at moving large quantities of heat and contaminants out of a structure. The technique and equipment gradually evolved over a number of years as firefighters who were not satisfied with traditional ventilation methods experimented to find a better way.

The Origins of PPV

The LAFD can be credited with being one of the first departments in the United States, if not the first in the world, to recognize the need to change the methods of ventilation for the benefit and safety of firefighters. Like many other departments in the country in the 1950s, after a fire was knocked down, they used electric exhaust fans to create a negative pressure in a structure to draw the toxic atmosphere outside. Over time they noticed that the air they were drawing inside only served to dilute the interior atmosphere, and little of the products of combustion were really being removed.

Turning the fans around

In 1954, firefighters began turning one electric exhaust fan into the building to push air inside the building rather than trying to pull all the air to the outside. Utilizing one fan did little to affect the interior atmosphere. However, when two fans were turned to blow fresh air into the structure, they moved so much smoke and toxic atmosphere out of the structure that it was readily visible. Firefighters were so impressed that they added a *third* electric fan. It overloaded the generator and all the fans stopped operating.[1]

LAFD's experience with the electric fans was the first sign that a new concept in ventilation was on the horizon, but equipment was needed to overcome the limitations of the ⅓-hp electric fans. They needed a reliable power source at the fire that could supply multiple fans. It soon became obvious that another type of fan would be required—one that had more horsepower and could operate without an electric power source.

As luck would have it, one of the fire captains interested in this new "reversed" method of ventilation happened to live next door to Roy McBride, who owned a company called Controlled Airstreams. This company produced gasoline-powered fans used in the utility business. The two collaborated on an experimental project to determine if gasoline-powered fans would be suitable in fire ventilation. They repeatedly built fires in metal trashcans inside their garages and then experimented with blowing out the smoke. The large front opening of the garage was used as the exhaust opening, and the smaller side entry door was used as the ventilation opening where the fan was positioned.

They observed a 3-hp gasoline fan (blower) clear a garage full of smoke. They also determined that a shrouded cage and a blade engineered for more efficient air movement increased the blower's volume and dynamics. This in turn increased the blower's ability to remove smoke from the inside of a structure.

After these experiments demonstrated the value of the new method, Chief John Mittendorf did additional research and developed and implemented the necessary training and standard operating procedures that would be used by the LAFD. The term *positive pressure ventilation* was first used in Los Angeles to describe this method to pressurize structures during overhaul and in stairwells of high-rise buildings.

Building a better blower

The new method worked well, but it could be better. One of the problems was that the only way to move air was to use fans designed for utility work rather than firefighting. In 1987, that changed as a result of the chance meeting of two hot air balloonists and a firehouse cook.

The two balloonists, Dexter Coffman and Don Hamman, were searching for a more efficient way to inflate a hot air balloon (fig. 4–1). Because there were no commercial blowers available for this, hot air balloonists used lawn mower engines with wooden blades attached in a frame enclosed by wire. But this type of blower had the bad habit of not wanting to stay put during operation and required that a member of the ground crew hang onto it.

Fig. 4–1 Two balloonists, Dexter Coffman and Don Hamman, looking for a better way to inflate a hot air balloon, developed a blower that caught the interest of firefighters experimenting with PPV. Courtesy Tempest Technology, Inc.

Hamman began experimenting with a blower that would do away with the extra crew member. He came up with a design that included a shock-absorbing engine mount, an aluminum blade, a shroud, and an aluminum impeller. This resulted in a more powerful blower design that was also safer for the ground crew to use.

At this point the fire department cook, who was Hamman's grandmother, entered the scene. She mentioned to Hamman that the firefighters she cooked meals for were experimenting with a new, radical ventilation technique called positive pressure ventilation. Soon after, the department began testing her grandson's redesigned blower on ventilation situations. They discovered that the new design, which incorporated a spun aluminum shroud and a seven-blade cast aluminum impeller, was superior to anything on the market. The addition of the shroud and the innovative blade created a "cone of air." This could completely cover the ventilation opening and produce enough air movement to significantly increase the effectiveness of PPV.

The success of these tests prompted Hamman to contact Coffman about starting a company to produce blowers specifically for fire service use and to promote the scientific study and use of this new technique. Tempest Technology Corporation was born of this effort. The company sponsored a national symposium that invited the most knowledgeable experts on ventilation practices in the country to discuss the new ventilation method. PPV started gaining wider acceptance in other fire departments.

PPV Fundamentals

Simply stated, PPV is the use of gasoline-powered blowers to force heat, smoke, and toxic gases out of an exhaust opening. Part of the beauty of PPV is its simplicity. Although it is very important that crews have adequate training in its use, it is not particularly difficult to explain, and it can be set up quickly on a fire scene. Before going into the details of using it on the fireground, some general information is necessary.

How PPV works

The science behind using gas-powered blowers to move heat, smoke, and toxic gases out of a structure is pretty straightforward. In a nutshell, air will move from an area of higher pressure to an area of lower pressure, utilizing the path of least resistance in search of equalization.

For example, one could consider what happens when someone breathes. During inhalation, the chest increases in size and creates a lower pressure area in the lungs. The air outside the body, which now has greater air pressure, flows into the area of lower pressure in the lungs. During exhalation, the chest decreases in size. Air moves out of the lungs, now an area of greater pressure due to contraction of the chest, to outside the body, where the pressure is lower. This process follows the laws of physics and is inviolable.

When a blower is positioned outside a door and the airstream is directed into the building, the air that comes through the blower housing as well as the entrained air around the blower increase the quantity of air inside the structure. Thus, the pressure inside the building is increased. When the pressure inside the structure becomes higher than the atmospheric pressure outside, air from the inside will exhaust from any opening other than the opening the blower is in, under normal conditions (fig. 4–2). (Note: It is useful in some situations to adjust the blower position so it will pressurize the interior and exhaust from the same opening. This variation will be discussed later.)

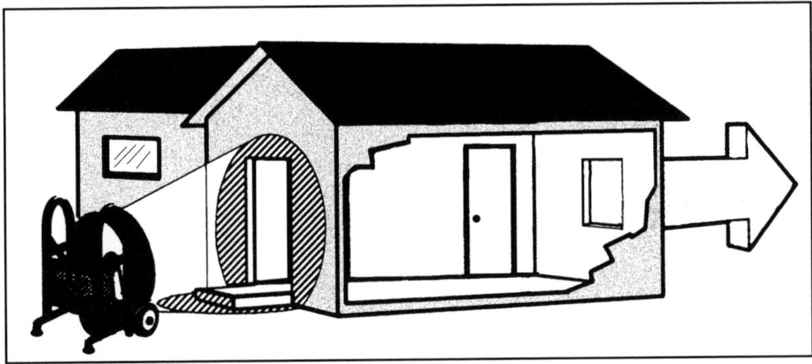

Fig. 4–2 Positioning a blower to force air into an opening increases the air pressure inside a building. Under normal conditions, this causes the atmosphere inside the building to exhaust out openings to the outside other than the one the blower is in. Courtesy Tempest Technology, Inc.

Positioning the blower so the cone of air completely covers or "seals" the opening and keeping the opening clear ensure that the building receives maximum pressurization. The authors have studied the balance between air volume exiting from a test space, efficient access, and other factors for more than a decade. They have concluded that positioning a blower 8 ft to 10 ft from an opening is ideal. A seal is optimum but not absolutely necessary if, due to conditions around the ventilation point, ideal positioning cannot be accomplished. Having to place the blower closer than the recommended 8 ft to 10 ft from the ventilation point only slightly decreases the total volume of the interior atmosphere that is exhausted (fig. 4–3).

Fig. 4–3 Positioning the blower 8 ft to 10 ft from an opening so the cone of air covers the opening ensures maximum pressurization and air entrainment. If ideal positioning is not possible, a wide range of variations will still produce adequate ventilation. Courtesy Tempest Technology, Inc.

Effective PPV raises the pressure of the interior atmosphere only slightly—just 0.1 to 0.2 pounds per square inch (psi). This pressurized air then exits through an exhaust opening of the fire's choice or one created by on-scene crews to direct the interior contaminants to the outside (fig. 4–4). This is a vast improvement over traditional roof operations that depend on the natural tendency of heated air to rise. This is especially true considering that when hose streams cool products of combustion, they do not rise as well. When compared to a standard-sized, 4-ft by 4-ft hole, the use of blowers increases the volume of air movement about 20 times, according to tests by the authors.

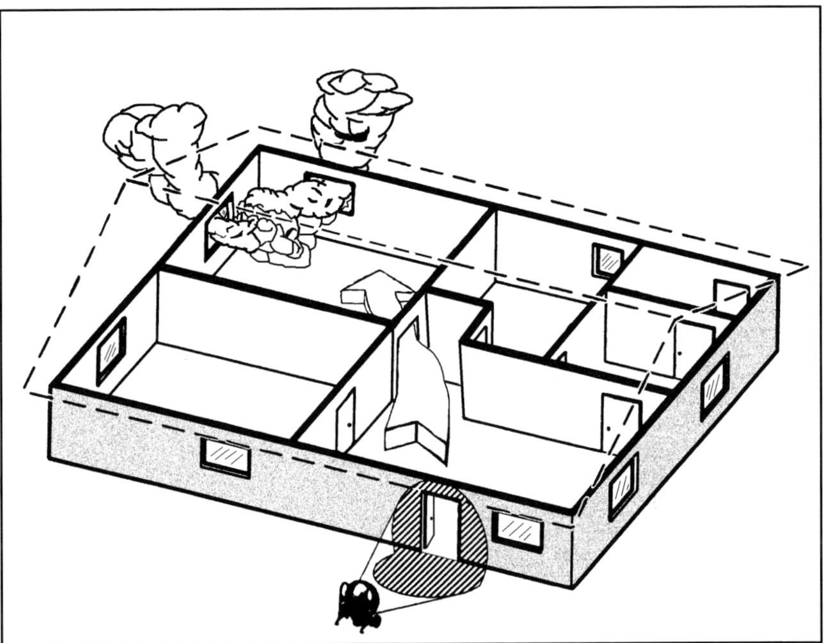

Fig. 4–4 Pressurized air exits through any exhaust opening of either the fire's choice or one created by crews to direct contaminants to the outside. Courtesy Tempest Technology, Inc.

Anatomy of a blower

One of the great advantages of PPV is the amount of equipment required. The list begins and ends with the blower.

A blower is a specialized piece of equipment. It must be mobile, durable, reliable, and easy to operate for it to be effective for fire service use. A shopping list of features is included here (fig. 4–5):

1. It should be gasoline powered. This makes them self-contained in terms of not requiring an external power source such as a generator. Electric or water-driven blowers take longer to set up, which makes them more difficult to coordinate with fire attack. There is a place for these blowers, though, which will be discussed later.

2. The engine should have as few switches as possible to operate the blower. Some blowers may have as many as four separate switches and valves to turn them on. A blower with a single choke switch is ideal. Simplicity is important.

3. The turn-off switch should return to the start position automatically. This eliminates another possibility for error and adds reliability.

4. It should be able to tilt. The ability to angle the airstream up or down makes a blower more adaptable to different situations on the fireground.

5. It should have an output capacity of 15,000 to 20,000 cfm. At this writing, blowers are rated by how much volume they can put into a test chamber. A new rating standard by ANSI (American National Standards Institute)/AMCA (Air Movement Control Association), Number 240–96, rates blowers on how much volume they can remove from a test chamber. As clearing the interior of the building is the primary goal, the new ratings better reflect their effectiveness. Tests are conducted in a constant chamber located at the AMCA testing facility in Illinois.

6. The blades should be large diameter. The larger the blade, the greater the volume of air moved. Of primary importance is the volume of the output (cubic feet of air moved per minute), not the velocity (distance per time unit). Multiple, single-piece aluminum blade systems generally provide the best rating for PPV. These blades also have a good safety record. Other blade materials have had issues with delamination, fractures, and total blade failure.

7. The blades should be surrounded by a solid shroud or cover. A solid shroud protects the blades and improves the dynamics so the blower can push a large volume of air.

8. It should be relatively lightweight. The blower should be light enough to be maneuvered by one firefighter, and it should have an attached handle to make moving it easy.

9. The engine should have overhead valves. Overhead valves decrease the amount of transitory carbon monoxide the blower generates.

10. It should have pneumatic tires. Wheels are a necessity, and pneumatic tires make it easier to move the blower up and down stairs and over terrain that hard tires would not negotiate easily.

11. Exhaust extensions are not important. According to tests conducted by the authors, exhaust extensions provide negligible benefit during PPA. After an interior attack is made, CO levels inside the structure can be more than 1,000 ppm. PPV reduces this to safe operating levels within a few minutes. The added set-up time, expense, and burn hazard from hot extensions are not worth the less than 5% decrease in interior CO they will produce. If reduced CO is necessary, an electric blower is a better option.

Fig. 4–5 Qualities to look for in a blower include: simple operation; an off switch that automatically returns to the start position; the ability to tilt; output of 15,000 to 20,000 cfm; large diameter blades; solid shroud; relatively lightweight; overhead valve engine; and pneumatic tires. Courtesy Kriss Garcia

Positive pressure terminology

An understanding of several important terms will be helpful to this discussion (fig. 4–6).

- **Blower.** A compact, gasoline-powered, high-volume fan specially designed for fire department use that, when correctly positioned outside a building, is capable of forcing air inside to pressurize the interior. Fire service blowers have wheels to aid in their mobility at a fire scene.
- **Ventilation opening/ventilation point.** The door, window, or other opening in a structure where the blower is placed to blow fresh air inside to pressurize the interior. Air from the blower enters the structure at this opening. In most circumstances, the best location is where interior crews enter the building.
- **Exhaust opening/exhaust point.** Any opening, which could be a doorway, window, attic opening, or an opening the fire makes, where the products of combustion exit the building.
- **Cone of air.** The pattern of the airstream created by a blower. The narrowest point of the cone is at the blower, and the cone gets wider as the distance from the blower increases.
- **Seal an opening.** Positioning a blower so the cone of air completely covers the ventilation opening.

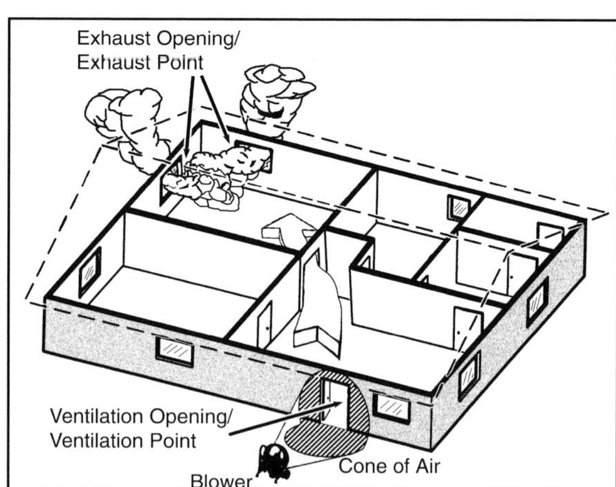

Fig. 4–6 Positive pressure terminology. Graphic courtesy Tempest Technology, Inc.

Setting up PPV at the Fire Scene

In its most basic form, two simple operations are required for PPV: (1) set up the blower at the ventilation opening; and (2) ensure an exhaust opening. As with any fireground operation, proper use of PPV requires that all personnel understand its operating procedures, practice its use, and learn its limitations. Firefighters should not turn an operating blower into a building until there is an adequate exhaust opening.

The ventilation opening

To accomplish PPV, a blower is set up outside an opening to the building, which is the ventilation opening. The door that crews are using to access the building is preferred for a ventilation opening. To maximize the flow of air, it is important to keep obstructions, including firefighters, from blocking this entrance.

Optimum blower placement depends on the size of the entrance opening and the size of the blower. Generally speaking, however, positioning the blower 8 ft to 10 ft away from the door is usually optimal with a 15,000-cfm to 20,000-cfm blower. The blower can be tilted back about 20° to 30° to direct the center of the cone of air at the vertical center of the opening. Also, multiple blowers may be used to increase the volume of air introduced.

The exhaust opening

Before the airstream from the blower is directed into the structure, it is important to ensure that there is an adequately sized exhaust opening. Adequate size means in the range of one to three times the size of the ventilation opening. This size range will efficiently provide the maximum ventilation in a timely manner.

Directing the airstream of a blower into a ventilation opening increases the pressure inside the building. Other than near the ventilation point and exhaust point(s), this increased pressure will tend to be consistent throughout the building—at the ceiling and floor and in the corners. When pressure is increased inside the building, any opening to the outside becomes a path for the interior atmosphere to escape outside to where the pressure is lower, much the same as air escapes from a hole in a car tire (fig. 4–7).

Fig. 4–7 When pressure is increased inside a building, any opening to the outside will provide a path for the interior atmosphere to escape to the outside, where the pressure is lower. Courtesy Ray Schelble

In the car tire example, air stops flowing out when the pressure inside the tire equalizes with the atmospheric pressure outside. In PPV, the working blower keeps the atmospheric pressure inside the building elevated, which continuously displaces the interior atmosphere to the outside. This airflow will quickly remove lethal heat and contaminants from inside the structure to the exterior, replacing the interior atmosphere with cool, fresh air. The effect will be maintained as long as the blower is in operation.

Ideally, in a structure involved in fire, the pressurized interior atmosphere flows through one main exhaust opening located near the fire area. Gases and heat will be efficiently removed in short order. This will be discussed in more detail later.

An important reason to position the blower in the entrance that crews are using to access the building is to ensure that it will not become an exhaust opening. At best, this would hinder crews with the heat and smoke that is being exhausted. At worst, it could become a dangerous situation.

Safety and precautions

Although PPV is safe for use in a wide range of situations, there are circumstances when it can cause or contribute to serious problems. For instance, PPV can mask the signs of hidden fire, especially after the fire is knocked down or during overhaul. The substantial air movement from PPV can minimize or completely hide indicators such as drift smoke and walls that are hot to the touch. Blowers should be turned off during the final stages of overhaul as crews are evaluating whether additional overhaul is necessary. Turning off blowers for approximately 10 minutes after the interior is cleared can help the overhaul crews find these problem areas and take care of them (fig. 4–8).

Fig. 4–8 PPV can mask signs of hidden fire. In addition to opening up all areas that have been exposed to fire, it is important to turn off blowers for about 10 minutes after the interior has cleared and do a thorough check for fire that may still be burning in hidden spaces. Courtesy Kriss Garcia

An exhaust opening becomes an exit for smoke and hot gases. Depending on the circumstances, the heat and products of combustion being exhausted can create problems for firefighters who may be between the fire and the exhaust opening. It can also make the area immediately outside the exhaust point untenable.

When PPV is used, the authors recommend temporarily removing all firefighters from the building after the fire is knocked down and before starting pressurization. This is just an added precaution to avoid putting firefighters at risk. Firefighters must remain clear of any potential exhaust openings and may start reentering through the ventilation opening 30 seconds after the blower is directed into the building.

Also, PPV (or any interior operation for that matter) must not be used when backdraft conditions exist in a building. The heat and gases inside the building must be safely vented before blowers can be used.

These and other safety issues will be more thoroughly discussed in chapter 6, which must be read and understood before attempting any operation involving positive pressure.

Fine-tuning PPV

One strong point of PPV is that it works well with a broad range of blower-positioning combinations and exhaust-opening locations and sizes. For the most effective ventilation, firefighters should try to maximize the exhaust-opening size and take advantage of external influences.

Balancing air in with air out. As stated earlier, the effectiveness of PPV directly depends on the volume of air moved. Older blowers were most efficient when the exhaust opening was controlled at between one and one and three-fourths times the size of the entrance opening. A newer, high-volume blower has no trouble maintaining pressurization with as many as three 20-ft^2 exhaust openings. The size of the exhaust opening for peak air movement depends on the number and capacity of the blowers in use.

A noticeable gasoline engine exhaust odor inside the ventilated area is generally an indication that the exhaust opening is not large enough. It may also indicate the blower is positioned too close to the opening. To improve performance, the size of the exhaust opening should be increased by opening another window or door in the same immediate area or by moving the blower farther away from the opening.

The effects of weather. Wind can have a dramatic effect on PPV, but only if it is blowing against the exhaust opening (fig. 4–9). Wind blowing against any of the other three sides will cause no appreciable loss in effectiveness. After a decade of aggressive experience with positive pressure, the authors have seen or heard of only a small percentage of PPV operations that required modification due to wind.

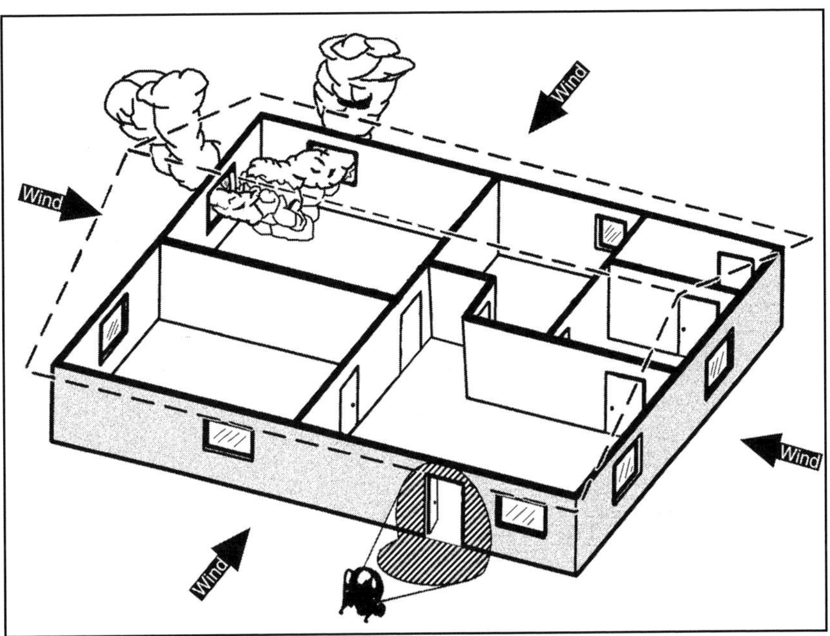

Fig. 4–9 Although few fires require that PPV be modified due to wind, problems can arise if a consistent wind is blowing against the exhaust opening. Graphic courtesy Tempest Technology, Inc.

Temperature, humidity, snow, and rain do not have a noticeable effect on PPV. Even though cold, damp conditions may not allow smoke to rise outside, these conditions will not limit the ability of blowers to move contaminants horizontally and, in most cases, vertically.

Blower Configurations

One blower will be all that is needed in the majority of situations. Configurations using multiple blowers can dramatically increase the volume of air moved and reduce the time needed to ventilate, especially in larger buildings.

Single blower

This is the basic setup and is most frequently used to ventilate the typical residential fire (fig. 4–10). The blower is placed about 8 ft to 10 ft from the opening. Under test conditions, blowers positioned at this distance have consistently moved the highest volume of air from test buildings.

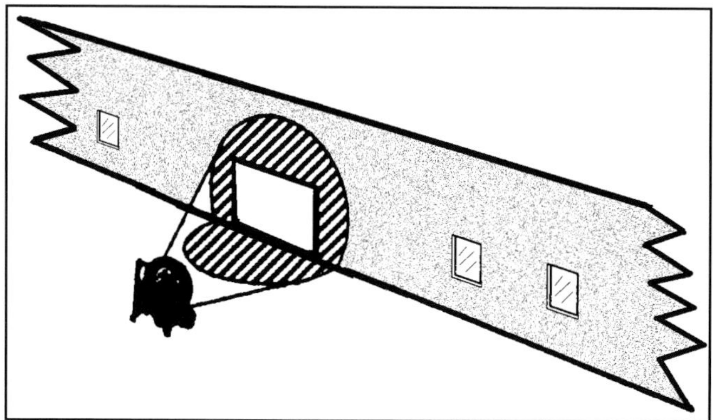

Fig. 4–10 Single blower. Courtesy Tempest Technology, Inc.

Firefighters already know that in some situations there will not be enough room to position the blower the recommended distance from the opening. Porches, landscaping, fences, stairways, and a wide variety of other obstructions can get in the way. The answer is to adapt. Firefighters should place the blower as close as possible to the recommended distance. The air volume exhausted from the structure may be reduced, but it will still provide very effective ventilation.

In-line (series) blowers

For enhanced effectiveness when moving air through a standard entrance opening (approximately 6½ ft high by 3 ft wide), two blowers should be placed in front of the door in-line with each other (fig. 4–11). One should be positioned about 2 ft to 3 ft from the door, and the other the recommended 8 ft to 10 ft from the door to cover the door with a pressurized cone of air. The closer blower provides a boost of pressurized air, while the farther one seals the opening and entrains additional air. It is important to always ensure enough room for firefighters to move in and out of the entrance.

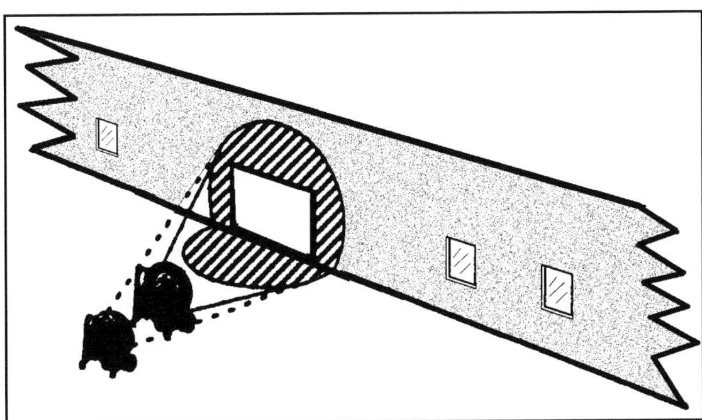

Fig. 4–11 Series blowers. Courtesy Tempest Technology, Inc.

When using multiple blowers with different capacities in-line, the authors recommend positioning the blower with the smaller capacity close to the opening to ensure that all its pressurized air enters the building. The larger blower should be positioned far enough behind the first to cover the entire opening with a cone of pressurized air. Positioning two blowers, with the larger one closer to the opening, creates a minimal reduction in overall operational efficiency. Tests demonstrate that adding a second blower in-line increases air volume by approximately 30%.

Parallel blowers

For openings wider than a standard-sized door, such as an overhead garage door, multiple blowers positioned parallel, or side by side, provide better coverage than in-line blowers (fig. 4–12). The number of blowers required and their positioning to completely cover the opening with combined cones of pressurized air will depend on the size of the door.

Some larger openings, a loading-dock door or a garage door, for instance, can be reduced in size by partially closing the door. The door can be blocked in the desired position by turning off the power, using Vice-Grips-style pliers clamped on the track below the door, or by using some other blocking arrangement that holds the door safely.

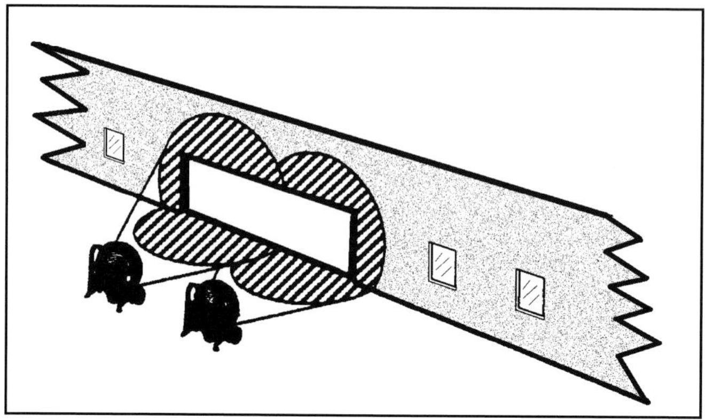

Fig. 4–12 Parallel blowers. Courtesy Tempest Technology, Inc.

Combination blowers

Multiple blowers can be set up in any number of configurations (fig. 4–13). Depending on the number of blowers available, a large area may be better ventilated by combining parallel (to improve coverage of the opening) and in-line (to increase air volume) positioning.

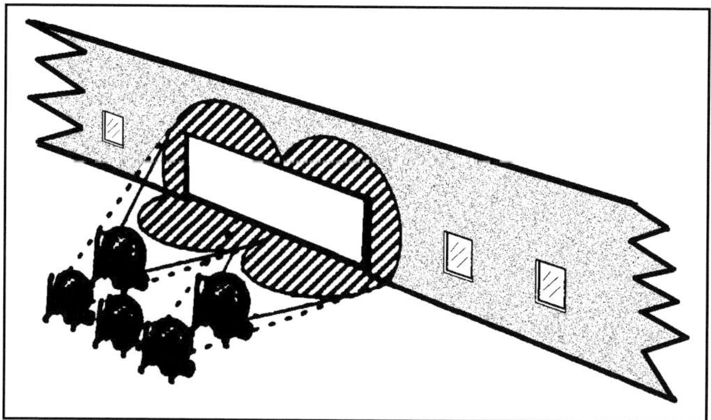

Fig. 4–13 Combination blowers. Courtesy Tempest Technology, Inc.

In some situations, one or more blowers must be positioned in the interior of a building to help direct the air movement from the outside blower(s) in a beneficial direction. Configuring multiple blowers this way can be of use in a building with sealed windows where the ventilation point and exhaust point may be through stairwells, such as in high-rises. It requires imagination and ingenuity on the part of the firefighters. Combination blowers are also effective in buildings where internal partitions divide the interior into areas that can be ventilated in sequence. This will be covered further in chapter 7.

V-point blowers

In this configuration, two blowers are positioned parallel to and several feet away from each other and aimed at the same opening (fig. 4–14). Each blower's cone of air is centered on the nearest edge of the door frame, so their airstreams intersect at the ventilation opening. Ideally, the angle for each blower should be about 45°. If necessary, one blower can be directed to seal the top of the opening and the other to seal the bottom. The V-point configuration can be used to supplement a single blower when another company arrives at the scene with another blower. It could also be used initially if enough blowers are available.

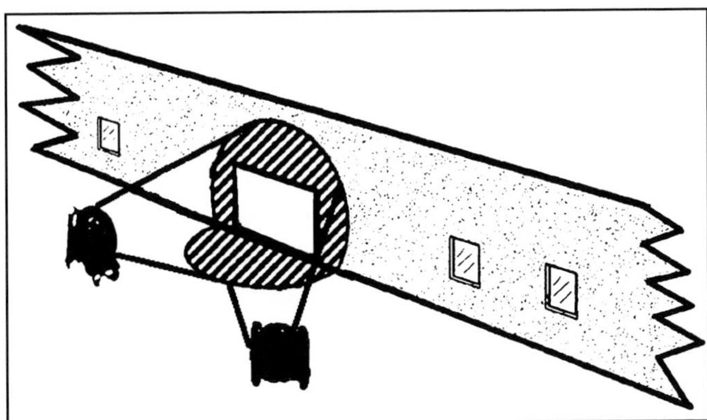

Fig. 4–14 V-point blowers. Graphic courtesy Tempest Technology, Inc.

This is a relatively recently developed blower arrangement that has two important benefits over a single blower:

1. It moves the blowers to the side of the opening, thereby keeping it clear for firefighters.
2. It increases effectiveness by combining the volume of two blowers.

V-point blowers provide the optimum airflow that two blowers can deliver. In tests by the authors, the V-point configuration delivered approximately 10% more exiting air volume than in-line, series, or parallel positioning.

Stacked blowers

When the ventilation point is a high vertical opening, blowers may be stacked on one another. Some blowers are designed to be stacked this way, but most are not. Care must be taken to secure blowers so they will not fall and injure someone or cause property damage.

Structural Influences

The type of building construction and occupancy has a great influence on how to best accomplish effective PPV. The subjects in this section will be addressed in more detail later in the book.

Areas with limited access

PPV can work effectively in an area that has only one door and no windows by using multiple blowers. First, a blower should be positioned close enough to cover only the bottom half of the doorway. Then, a second blower should be positioned at a right angle to the first blower to provide a flow of air across the top of the doorway. The first blower forces fresh air into the bottom of the opening, which forces contaminants out of the top of the opening. The second blower sweeps away contaminants without pushing them back into the airstream of the first blower (fig. 4–15).

Fig. 4–15 To ventilate a room with one opening, the blower should be positioned close enough to cover only the bottom half of the doorway. A second blower can be positioned at a right angle to the first and tilted upward to clear contaminants as they exhaust from the top of the opening. Courtesy Tempest Technology, Inc.

Basements

A basement accessed by a door inside a building can be pressurized by placing a blower at the main entrance to the building and closing off the other areas of the building (fig. 4–16). Putting an additional blower inside the building at a basement entrance will work, as well. Windows or other outside entrances should be used for exhaust openings. Because of the limited number of exterior openings in many basements, conditions may warrant taking out windows away from the immediate fire area to provide enough exhaust.

When a basement has only one entrance and no windows, firefighters can adapt the procedure outlined in the previous section, "Areas with limited access."

Fig. 4–16 Basements can be vented by putting a blower in front of the entrance and opening basement windows. If the basement entrance is inside like this one, closing off the rest of the building except for one opening to provide outside air for the operation will direct pressurization into the basement. Courtesy TCVFD

Multiple-story dwellings

In a multiple-story dwelling without a built-in smoke removal system, firefighters should start at the lowest level and ventilate toward the top floor to exploit the tendency of hot smoke and gases to rise. They should position a blower at an appropriate entrance opening and employ a technique called sequential ventilation.

To sequentially ventilate a building, each floor is isolated either by shutting the stairwell doors to other floors or by closing off all windows and exhaust openings outside the area to be ventilated (fig. 4–17). Only windows on the floor to be ventilated should be opened, and they should be closed when the floor is cleared. Progress should continue upward this way until each floor has been cleared of smoke.

Fig. 4–17 Multiple story buildings can be vented sequentially by isolating rooms, floors, or other smaller areas and providing an exhaust opening for each. Courtesy Kriss Garcia

By controlling the exhaust openings that are open at any one time and proceeding in sequence, one blower at the ventilation point may be able to supply enough pressurization to ventilate a sizeable building. As long as the blower is operating at the ventilation point, the floors without exhaust openings will still be pressurized. However, air will only be moving along a path from the entrance opening through the exhaust openings. Sequential ventilation can be used in the same way to horizontally or vertically ventilate any building with compartments that can be closed off. Sequential ventilation is discussed further in chapter 7.

Commercial buildings

Commercial buildings come in all shapes, sizes, heights, and types of occupancy. The following should be considered when planning ventilation operations for commercial structures.

Lightweight construction. Since the 1960s, lightweight construction methods have become increasingly popular in all types of commercial buildings. As was discussed earlier, after 5 to 10 minutes of exposure to fire, roofs with lightweight supports will not even support their own weight (fig. 4–18). The added weight of crews performing vertical ventilation could quickly have disastrous consequences. Also as discussed previously, vertical ventilation is generally not effective in support of initial fire attack the majority of the time. This can be due to complications or simply because of the increased time it takes to complete. PPV is the best way to ensure firefighter safety and effective ventilation in these buildings.

Fig. 4–18 After 5 to 10 minutes of fire, roofs with lightweight supports will not even support their own weight. Courtesy Kriss Garcia

Large open areas. Commercial buildings such as warehouses and manufacturing occupancies may have large open areas that require a large volume of air movement to remove contaminants. These occupancies require multiple parallel blowers capable of providing adequate volumes of pressurized air (fig. 4–19). Multiple blowers can be used to pressurize large entrance openings, and doors can be partially closed if necessary to better pressurize the entrance opening (fig. 4–20).

Fig. 4–19 Adequate volume to ventilate larger buildings requires an appropriate configuration of multiple blowers. Courtesy Kriss Garcia

Fig. 4–20 Partly closing an overhead door can create a manageable ventilation opening. Crews also elevated this blower on a stack of pallets to provide the best coverage of the opening. It is important to always ensure that a blower is positioned securely and cannot fall. Courtesy Ray Schelble

Compartmentalization. Large structures that are divided into smaller areas such as stockrooms, workstations, and offices should be sequentially ventilated in a preplanned, coordinated operation. Partitions for these interior areas can be looked at as if they were exterior walls. If they do not have openings to the outside, these smaller areas should be vented out to the main area of the building, which should have an operating ventilation point and exhaust point. Whenever possible, firefighters should reduce large areas into smaller ones and ventilate sequentially for greatest efficiency.

Unintended exhaust openings

Large ventilators, broken skylights, or ventilation openings created by fire or firefighting operations can negatively impact PPV. Doors and windows should be closed off to maximize the flow of pressurized air to direct products of combustion through these vertical openings (fig. 4–21).

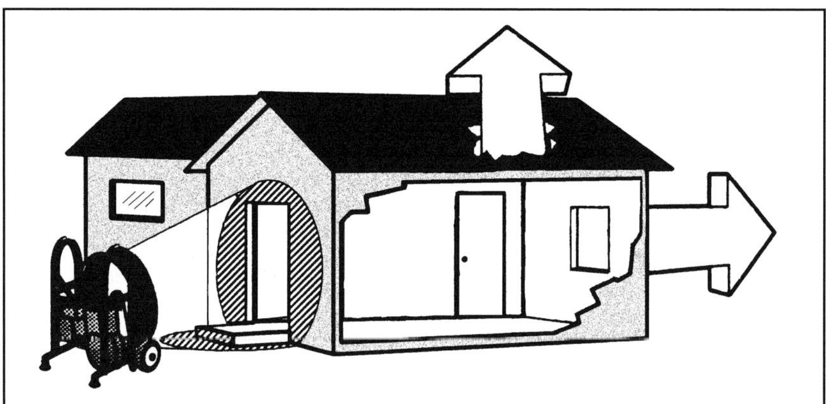

Fig. 4–21 PPV will be effective no matter where exhaust openings are located, although the openings may need to be controlled to maximize ventilation if openings are extensive. Courtesy Tempest Technology, Inc.

Of course, a fire that has burned through a roof limits options with any ventilation method. A hole in the roof remains a hole in the roof and cannot be altered when conditions call for a change in tactics.

No exhaust opening

The airstream generated by the blowers increases the pressure of the atmosphere within the fire building, creating conditions for it to exhaust to an area of lower pressure given the opportunity (an exhaust opening). When the entire ventilation opening is pressurized without having an exhaust opening, the pressurized interior atmosphere cannot escape. Under these conditions, the building interior maintains the highest possible pressure. This concept will become important later when discussing positive pressure to protect exposures.

Controlled Ventilation

PPV can be effectively harnessed by firefighters to selectively direct the flow of heat, smoke, and products of combustion to any area inside or outside an intact building (fig. 4–22). Simply opening and/or closing doors or windows can alter the ventilation to achieve a desired effect. Air volume can be increased by adding more blowers. It can be stopped at will by turning off the blowers when interior crews need to check for hidden fire or if it is not producing the intended results (fig. 4–23).

Fig. 4–22 PPV can be readily adapted to conditions encountered on the fireground. Courtesy Dan Walker

Fig. 4–23 Unlike a hole in a roof that becomes a permanent feature for the duration of a fire, PPV can be altered to achieve a desired effect, increased by adding more blowers, or stopped completely by the flip of a switch. Courtesy Kriss Garcia

Part of the adjustment firefighters go through when learning PPV is remaining flexible in their thinking to take advantage of the options PPV can offer. Incident commanders can reevaluate their action plans and make adjustments. No other method of ventilation allows for this combination of effectiveness, timeliness, and flexibility.

Pressurization is about altering the environment in which firefighters are operating or the environment in which someone is trapped. With pressurization, firefighters have more control over the environment to assist in safe operations.

The next chapter explains how PPV combined with fire attack can achieve the ideal of ventilation coordinated with fire attack.

Reference

1. Coffman, Dexter. 1999. Personal correspondence to Bill Manning (editor, *Fire Engineering Magazine*), September 25.

5

PPV + Initial Fire Attack = PPA = Coordinated Attack

The concept of ventilation coordinated with fire attack discussed in chapter 1 has been referred to in firefighting textbooks and trade journals. On close examination, however, there really has not been much put on paper that addresses how this coordinated attack can be accomplished in a manner that benefits the victims and the firefighters. The same goes for issues involving lightweight building construction that is not designed to hold up its own weight in a fire, much less that of fire crews.

These texts contain often-repeated statements of how ventilation and fire attack crews must communicate with each other and how the IC needs to coordinate these functions. There are repeated warnings of early collapse and fast fire spread. All of this is good advice, but there really has not been much in terms of how to actually accomplish these tasks. Unfortunately, firefighters have still been limited to using the same conventional methods to ventilate that, depending upon the department, often involve sending crews to the roof.

This chapter is where it all comes together. It is the nuts and bolts of how departments can achieve the ideal of systematic, coordinated, life- and property-saving ventilation without putting crews or victims in harm's way. This is where ventilation achieves coordination with fire attack—positive pressure attack.

As with the previous chapter, the reader should keep in mind that precautions must be observed for safety and effectiveness. PPV or PPA should not be attempted until firefighters read and understand chapters 4, 5, and 6, and until all personnel are trained in its use.

Taking PPV a Step Further

In 1987, while Coffman and Hamman were researching a more efficient way to inflate a hot air balloon, PPV research was also being conducted elsewhere. In Kern County, California, fire chief Cliff Allmon was experimenting and testing various blowers trying to duplicate the positive pressure techniques that were being used by the LAFD (fig. 5–1). This was probably one of the first documented trials using fans to remove smoke from a structure, although artificial smoke was used for the trials. They filled the firehouse dormitory with smoke and then used three different types of fans or blowers to determine which one would clear the dormitory the fastest. They placed the fan at the front door of the dormitory and opened a window in the dormitory for an exhaust opening. It took 30 minutes for the electric fans to begin clearing the smoke. A 3½-hp gas blower needed only 10 minutes to completely clear the room.

Fig. 5–1 The LAFD is credited with developing practical applications of PPV in the 1950s and 1960s, about the time this 1958 Seagrave was new. In the late 1980s in California, Chief Cliff Allmon of the Kern County Fire Department began experimenting with the PPV techniques the LAFD had developed. Courtesy LAFD Museum

They also tested a 5-hp blower that had been originally designed to inflate hot air balloons. This one cleared the smoke out of the area in only three minutes. The Kern County Fire Department decided to buy the more powerful and effective 5-hp blowers and to use them to remove the products of combustion from structures after the fire had been extinguished.

The next logical question

Chief Allmon then asked the next logical question, "If we can clear the structure this fast after the fire, why can't we do the same when it is in the free burning stage?" It was an excellent question.

The reaction he got was the standard "look at what happens when you blow into a campfire" response. In other words, if someone adds oxygen by blowing massive amounts of air into a burning structure, the result will be to "fan the flames" and spread the fire. Chief Allmon was not convinced they were right.

More firefighters were starting to realize that PPV was a great improvement over vertical or negative pressure methods. Its use grew for post-knockdown ventilation. Then, Chief Allmon responded on a nighttime house fire that involved heavy amounts of smoke and reports of trapped victims inside. Because of the urgent need to make a fast rescue, he ordered the new blowers be set up shortly after arrival to clear the house of heat, toxic gases, and smoke.

Three children were rescued and, in fact, were reported to be reviving on the couch when firefighters found them. This was possible because of a courageous order given by a smart, innovative fire chief. Because of his study and experimentation with PPV, Chief Allmon had the foresight to understand how the new method could truly benefit the fire service. He subsequently began live fire exercises in structures to better understand the principles of PPV and to develop the procedures for its use as an aggressive fire attack tool.

The beginnings of PPA

The major contribution PPV made to firefighting at this point was to make the interior environment more tenable during overhaul and salvage. In 1989, authors Reinhard Kauffmann and Kriss Garcia were attending a National Fire Academy class at Utah Valley Community College in Provo, Utah. This was done in order to improve their chances of being promoted to lieutenants on the Salt Lake City Fire Department. As fate would have it, the instructor was none other than Chief Allmon. To this day, neither of them can remember what the name of the course was or what they should have learned. The reason is that Chief Allmon began discussing the new techniques he was developing in his department involving the use of gas-powered blowers to ventilate structures involved in fire. They recall spending most of the class time discussing this new technique and then personally talking to him about it.

After the class concluded, the authors briefly discussed PPV and what they understood about it. Then they put it on the back burner to concentrate on the upcoming promotional examination.

A fatal house fire

Both were promoted in 1990. On a cold December night that same year, Reinhard was working as the officer on Salt Lake City's Rescue 8, an advanced life support (paramedic) engine. He tells the following story:

> *We were assigned to assist at a house fire with a known trapped child. A babysitter was at the home when the fire broke out and reported to first arriving companies that she could not account for the child. Neighbors were already trying to locate the parents, who were attending a nearby church function. Rescue 8 was assigned to perform search and rescue on the second floor. I recall moving toward the front door, from which heavy smoke was billowing, and noticing that the truck crew was placing a ground ladder to the building to open up the snow-covered roof for ventilation.*
>
> *We navigated up the stairs to the second floor through heat and smoke and almost zero visibility, and turned left toward what we thought were the bedrooms. At this point we found the small boy and fought our way back down the stairs to the front door.*
>
> *As we left the smoke-charged house, I noticed a ground ladder to my left and then heard the sound of a power saw being started. There was a waiting ambulance in the street. As we moved toward it with the child, I saw two people, who had to have been the parents of the victim, looking at me. They were clinging to each other, crying hysterically and horribly distraught, as we quickly moved past them. To this day I remember that as much as I wanted to, I could not make eye contact with them, because I had a gut feeling that their small boy was not going to survive.*

> *Paramedics were waiting in the ambulance to begin advanced life support, which included placing a tube down the little boy's airway. When they opened his tiny, lifeless mouth and began placing the tube, they noticed a terrible sight. He was full of thick, black particles of soot and had suffered severe burns to his respiratory tract.*
>
> *As the lights and siren of the ambulance faded down the street, I felt that we, as firefighters, had failed this family in our most important calling—that of saving life. As I walked back to the apparatus with the crew, I looked up and noticed that the roof was finally being opened. It was only 10 minutes, but it seemed like an eternity.*

When Kaufmann and Garcia teach their classes, they routinely ask students a series of questions. The first is, "How many of you have made rescues?" The second is, "How many of you have made survivable rescues of a victim who was in the area compromised by fire after you applied water?" In this case, *survivable* is defined as a victim who eventually was discharged from the hospital and survived 30 days. The findings from this informal survey put the survival rate of these rescued victims at under 2%.

The pieces come together

Reinhard continues his story:

> *After a few days had passed, I called Kriss and reviewed our fireground operations, hoping maybe to find something that would avoid the same tragic outcome at a future house fire. As if in an epiphany, we suddenly both came to the realization of what Chief Cliff Allmon had given us: a way to quickly remove heat, gases, and smoke; a method that could be used by the first-arriving company at a fire; and a method that might have saved the small boy at that house fire.*

Over the next several years, the two started pulling all the pieces together. They started by reviewing what Chief Allmon had discussed, developing their own operating procedures and doing tests to substantiate ideas with facts (fig. 5–2). Many of their early tests and evaluations were aimed at proving the theory wrong. They knew that they would have to convince their coworkers with emotion and demonstrations rather than merely presenting statistics and facts on paper.

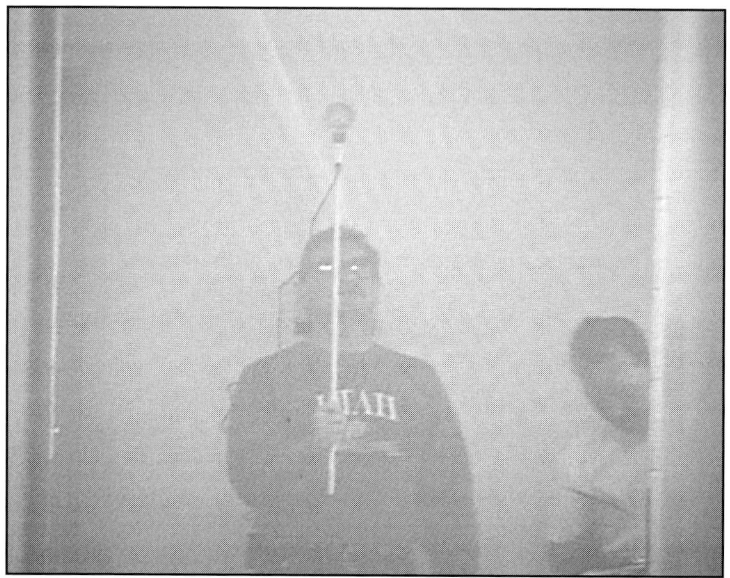

Fig. 5–2 In a still shot from a 1991 video, Reinhard checks the air velocity and Kriss records the readings in the artificial smoke inside the SLCFD training facility. Courtesy SLCFD

By the early 1990s, enough research had been done to justify using PPV, but trying to convince firefighters with pure facts and statistics was a complete failure. Chief Allmon's ideas were a step beyond that. Firefighters had to be convinced by real life, side-by-side comparisons and hands-on, live fire observation and experience.

To emphasize this new adaptation of PPV to fire attack, Kriss began calling the ventilation technique positive pressure attack, or PPA, to emphasize its important role in initial attack. Even though it is little more than timely PPV, the term PPA clearly distinguished its role in fire attack rather than post-knockdown.

PPA Fundamentals

During their study of PPV, the authors made an interesting discovery. They learned that when initial-arriving crews placed blowers into operation as they pulled their attack hoselines, ventilation suddenly became an effective component of a coordinated fire attack. Not only could positive pressure be put in operation immediately by the first-arriving crew without adding time to the operation, but it could also have a profoundly positive impact on rescue and fire attack.

Setup for PPA is basically the same as for PPV except for one very important difference. The blowers are set up before firefighters enter the structure and are operated in conjunction with the initial fire attack; thus the name positive pressure "attack." In other words, PPA is PPV started at the beginning of fire attack rather than after.

PPA is not a complicated undertaking, but attention to key procedures and precautions is absolutely necessary to ensure maximum efficiency and safety. These procedures allow a lot of latitude to adjust for conditions encountered and will still provide good results. The basic procedures include: positioning the blower, creating an exhaust opening, and beginning pressurization and fire attack.

Positioning the blower

On arriving at the scene, one firefighter takes the blower off the apparatus and wheels it to the ventilation opening (fig. 5–3). The preferred ventilation opening for initial fire attack is the entry that firefighters will be using themselves to enter the building. Ideally, the blower is placed about 8 ft to 10 ft from the ventilation opening. The blower should be started immediately. However, the airstream should not be directed into the ventilation opening yet. First the firefighter assigned to the exhaust opening (who also checks the building exterior for hazardous conditions) has to make sure there is an adequate exhaust opening.

Attack hoselines should be extended in such a manner that they will not move, interfere with, or tip over the blower during the attack. Simply put, firefighters must be careful not to tangle the hose around the blower.

Fig. 5–3 One crew member takes the blower from the apparatus and wheels it to the ventilation opening. Courtesy Kriss Garcia

As with PPV, in situations where there is not enough room to position the blower the recommended 8 ft to 10 ft away, it should be placed as close as possible to this distance (fig. 5–4). The volume of air entering the structure may be reduced, but the high-volume blowers will still ventilate effectively. If conditions allow blower placement only 4 ft from the opening instead of 8 ft, it is still far better than leaving the blower on the apparatus. The crew still must get the blower out and put it into operation! If they cannot meet the recommended blower placement, 3 ft to 4 ft away will still offer dramatic and positive results during fire attack.

An important point needs to be made here. It is imperative that a crew member on any first-arriving apparatus is always assigned to immediately take the blower from the apparatus and position it near the entrance. Making PPA an accepted part of normal operations requires systematic implementation.

Fig. 5–4 Ideally, the blower is placed 8 ft to 10 ft from the opening, although nearer or farther away will still provide effective ventilation. The blower is pointed away from the opening and the engine is started. Courtesy Kriss Garcia

Creating an exhaust opening

While one firefighter on the crew takes the blower to the fire building, a second (which could be the officer) takes an appropriate tool from the apparatus to create an adequate exhaust opening. (A 6-ft to 8-ft pike pole is ideal for a residence or small commercial building.) The second crew member then proceeds to the fire building to perform two important duties.

The first duty is to make sure there are no life hazards from a potential exhaust opening, such as somebody awaiting rescue at a window. After the building has been pressurized, any opening to the outside has the potential to vent large quantities of heat and other products of combustion. While this does not happen often, it can cause serious injury to anyone in or near the opening (fig. 5–5).

Fig. 5–5 After making a quick survey of the building exterior for life hazards from potential exhaust openings, and from a safe position, a crew member makes an exhaust opening and notifies the other firefighters. Courtesy Kriss Garcia

The second duty is to ensure there is an exhaust opening (fig. 5–6). This could be an existing opening, perhaps one the fire has made on its own, or a window can be broken out. The size of the exhaust opening should be substantial enough to allow as much fire and smoke to be removed from the structure as possible. Two times the area of the ventilation opening is ideal, but one to three times the size of the ventilation point is an acceptable range.

In most situations, this crew member must notify attack crews of potentially hazardous situations before ventilation is started. A possible exception would be certain conditions in an especially large building or buildings with difficult rear access that cannot be checked visually in a timely manner. Under these conditions, if it appears that heavy smoke and fire are obviously venting from a particular area of the building or the attic, and the exhaust point is not readily visible, firefighters can assume it has self-vented and make the attack. If there is heavy fire showing from the rear and a timely visual observation is not possible, they must assume that no one could survive those conditions if they are at the exhaust point. PPA should be started without delay. Later-arriving crews can pressurize and protect any potential exposures. (Chapter 6 offers further information on dealing with exposures.)

Fig. 5–6 Hazards at the exhaust opening include venting heat and products of combustion and flying glass. Safety and protective equipment are of the highest importance. Courtesy Kriss Garcia

The exhaust opening should be made as soon as possible, before the airstream from the blower is directed into the ventilation opening. It should be located at or as near as possible to the area that is involved in fire (fig. 5–7). When uncertain which area the fire is in, a best guess will do fine. If that means making more than one opening, firefighters should go ahead. Effective ventilation will still be achieved because of the high-volume blowers. The pressurization will be sufficient to remove smoke and heat.

Typically the exhaust opening will be a window or door. If a window or door is intact on arrival, it should be opened up completely, using a tool if necessary to break out glass or anything blocking the opening. Firefighters must be careful of flying glass fragments or shards that can slide down an axe handle and cut hands. If fire has already self-vented through a window or door, it is important to ensure that the opening is completely open. Firefighters can use a tool to take out the rest of the window or door and enlarge it as necessary.

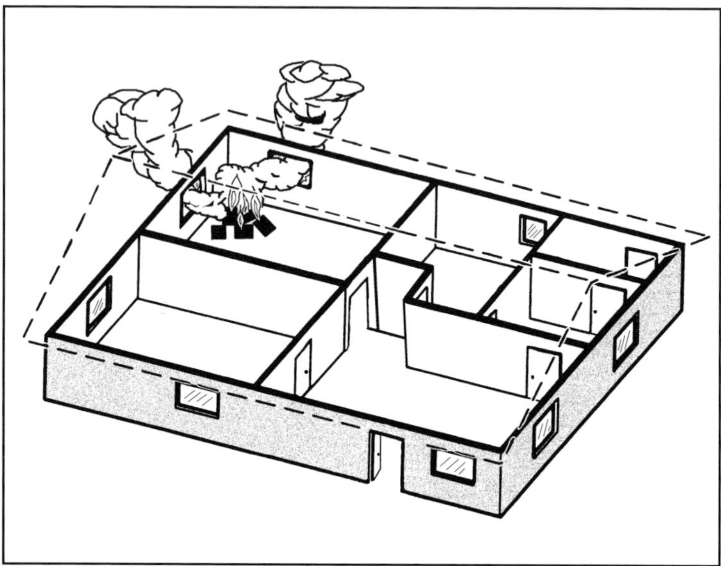

Fig. 5–7 The exhaust opening should be made in an area at or as near as possible to the area involved in fire. Courtesy Tempest Technology, Inc.

In addition to people awaiting rescue at windows, the exhaust opening presents hazards to people and property near the opening outside. When the blower is started, large quantities of smoke and possibly flames can forcibly exhaust. Again, firefighters should not begin ventilating if someone is at an open window awaiting rescue. There is also danger to anyone standing close to an exhaust opening when pressurization begins. Furthermore, there is also potential danger from choosing an exhaust opening where the heat and gases can impinge on an adjacent structure or object.

When making any exhaust opening, firefighters should open it from the side and move away.

Beginning pressurization and fire attack

As explained, the assigned crew member must check out the exterior of the building, verify or make an exhaust opening, and give the okay. If the attack line is charged and firefighters are ready at the nozzle, the blower should then be directed into the ventilation opening to begin pressurizing the building (figs. 5–8, 5–9, and 5–10). If using an older, smaller capacity blower (under 10,000 cfm), the firefighter assigned to the blower should check the seal around the top and both sides of the opening. Air movement felt approximately 6 in. to 12 in. outside the opening means there is optimum pressurization, although optimum pressurization is not necessary for the ventilation to still be effective. This step is not necessary with the new blowers that move 15,000 cfm or more. After the blower is directed into the entry, all personnel must stay out of the airstream as much as possible so airflow is not impeded.

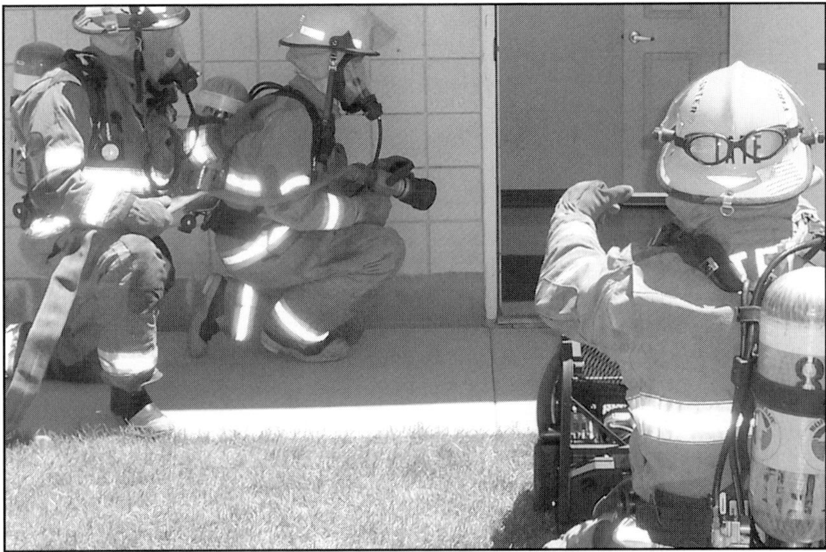

Fig. 5–8 After the assigned crew member has checked the exterior of the building, verified or secured an exhaust opening, and given the okay, the blower can be directed in the ventilation opening. Courtesy Kriss Garcia

After the blower is directed into the structure and operates for approximately 30 seconds, firefighters can enter the building for rescue and fire attack. This 30 seconds allows enough time for the blower to clear the interior ahead of the attack crews. This may seem like a long time, especially when there may be victims. But considering the time it would take to search by groping around in the smoke and heat, it really is not so long. After 30 seconds or so, interior conditions will rapidly improve, and crews will be able to enter the structure better able to see for search and rescue and to navigate to the seat of the fire in less time.

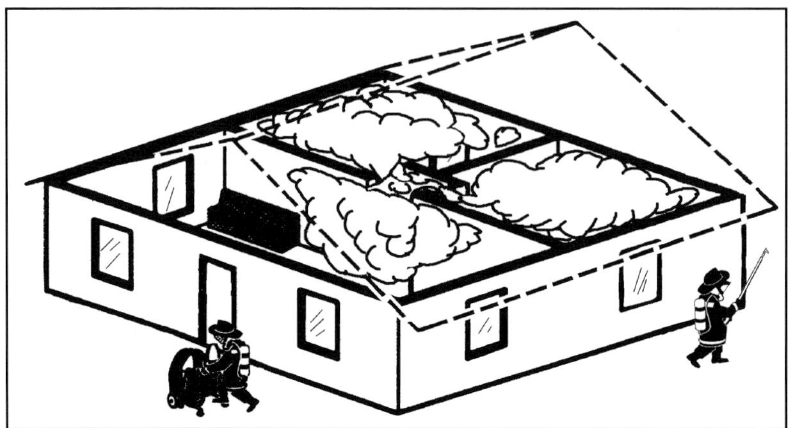

Fig. 5–9 Figures 5–9, 5–10, and 5–11 illustrate how PPA progresses. Here, one crew member takes the blower from the compartment, wheels it to the entry, and starts it without directing it into the building. A second crew member takes a pike pole or other suitable tool from the apparatus, surveys the outside of the building, and makes or improves an exhaust opening at or near the location of the fire. A third crew member pulls an attack line. The driver/operator readies the pump. Graphic courtesy Tempest Technology, Inc.

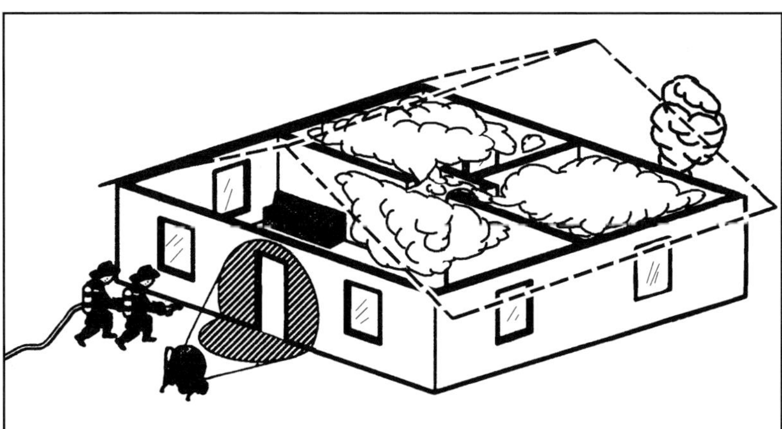

Fig. 5–10 Next, after getting an okay from the firefighter assigned to the exhaust opening, the blower is directed into the ventilation opening. Graphic courtesy Tempest Technology, Inc.

Conditions improve in an amazingly short time after directing the blower into the opening. After waiting 30 seconds, firefighters can usually walk upright through the structure because the temperatures decrease and visibility increases so quickly (fig. 5–11). Rather than enter after a set time interval, firefighters soon learn that they can sometimes enter the building by actually following the wall of smoke and heat as the pressurization forces it out the exhaust opening. One member of the attack crew should always take in a tool to check for hidden fire in the ceiling overhead before advancing too far into a structure.

Fig. 5–11 After blowers begin operating, interior conditions should improve rapidly, and entry can often be made standing upright. Graphic courtesy Tempest Technology, Inc.

The evolutions in chapter 9 detail specific tasks for each crew member when performing PPA.

Fine-tuning PPA

PPA is very effective under a wide range of conditions if firefighters pay attention to certain details and variations in its application.

An effective exhaust opening

Ideally, the exhaust opening should be located as close to the fire as possible and have no obstructions. It is important to make sure that products of combustion are exhausting from the opening when the blower has been directed into the structure. If nothing is exhausting from the opening, or interior conditions are obviously not improving rapidly, firefighters should move to another location and create another exhaust opening. Firefighters may have to open several windows or doors before getting the right one when the area involved in fire is difficult to determine. Also, interior doors may be closed, blocking off the airflow.

The size of the exhaust opening is not as limiting a factor with PPA as it is with more passive types of ventilation, since the interior atmosphere is being pressurized by the blower. Not only does the blower create pressure, but fire also increases the pressure due to expansion of the atmosphere due to heat. The amount of pressurization created by the fire is approximately 0.5 psi, or about 25% of the pressure created by a blower. This figure was measured at a roof ventilation hole with a fire burning and no blower operating. In this scenario, the hole was made in the optimum area with none of the compromises talked about earlier that limit the effectiveness of vertical ventilation. Pressure was then measured at the same point with a blower in operation.

Firefighters should remember that regardless of the size of the building, the exiting volume for a given blower will be the same from the same size opening. This is true whether in a single-family dwelling or in a warehouse. However, the percentage of the interior volume exhausted will be much more in a smaller structure than in a larger structure. In other words, with both the blower and the exhaust opening being equal in size, contaminants will be ventilated more efficiently from the smaller structure than from the larger. More volume must be moved to clear the larger building (fig. 5–12).

Opening an exhaust point in an area that is not near the main source of the fire does not "pull" fire into another area. This has been demonstrated repeatedly on multiple controlled structure burns and in specially designed props. The cooling effect from the large volume of outside air being introduced discourages the fire from spreading. It soon becomes obvious to interior crews when additional exhaust openings should be made.

The bottom line is that for the victim lying on the floor awaiting rescue, there is no bad way to go about replacing heat and smoke with cool, fresh air.

Fig. 5–12 Regardless of the size of the building, the volume exiting from the exhaust opening for a given blower will be the same from the same size openings, whether in a single-family dwelling or warehouse. The larger the area to clear, though, the more volume must be moved to effectively clear the interior. Courtesy TCVFD

Coordination with attack

For PPA to work, it must be approached in a systematic way. This is especially vital for the first-arriving fire company, because each crew member needs to perform assigned tasks. Each firefighter on the crew has a specific responsibility, and completion of each task is confirmed through either standard operating procedures (SOPs) or direct communication with the company officer. These tasks are not complicated, nor are they difficult, but they are important!

When implementing PPA on the fireground, the following information must be known by the officer and communicated to all firefighters working at the incident:

- The location and status of the exhaust opening
- Any potentially hazardous situations
- The location and status of the blower
- The position and status of the attack lines
- The source and status of the water supply

Once the exhaust opening is ensured, the airstream from the blower is directed into the ventilation point, and the building is pressurized. Then the interior attack crew makes their initial entry.

Readers should not let the simplicity of this fool them, however. To accomplish PPA, crews must be thoroughly trained and understand each task. To be coordinated, the operation has to run smoothly, and firefighters must follow the sequence. As stated earlier, PPA requires a coordinated and systematic approach. Training is the key to making it happen (fig. 5–13).

Fig. 5–13 Readers should not be fooled by the simplicity of PPA. To accomplish PPA safely and effectively, crews must be thoroughly trained and understand each task. Training is the key to making it happen. Courtesy Martha Ellis

Keys to successful PPA

There are three important keys to getting the most benefit from PPA as an attack option. First, the whole department must commit to it. A less-than-committed approach will not work. If this part sounds simple, it is not. Anything less than a full commitment is an invitation to fail. Implementing PPA on a fire department requires everyone on the department, from the senior fire officers to the firefighters on the street, to be willing to embrace the change. Departments must be willing to provide enough blowers and to properly train their firefighters.

Second, blowers must be readily available to the first-arriving crew, and there must be a system in place so firefighters will put them into operation when hoselines are stretched for the initial fire attack. Really, the only way that this can be done is to have blowers on every apparatus and to have all personnel trained (fig. 5–14).

Fig. 5–14 One of the keys to getting the most benefit from PPA is to have blowers readily available and taken to the building by the first-in crew. Blowers should be on every apparatus, and everyone on the department must be trained. Courtesy Kriss Garcia

Third, PPA should be used for aggressive interior attack on incidents where the initial-arriving crews can make a rapid entry into the area involved in fire. Large, complex fires require that all aspects of the situation be evaluated, and PPA may need to be deployed in a more methodical, less aggressive way.

PPA should be used at most fires that require ventilation. The authors rarely assign firefighters to roof operations for ventilation any more. Other departments can achieve the same success, but PPA must become the standard for fireground operations and not a specialized, rarely used form of ventilation (fig. 5–15). If not used as much as possible, proficiency at PPA will quickly decline, and the benefits will be lost.

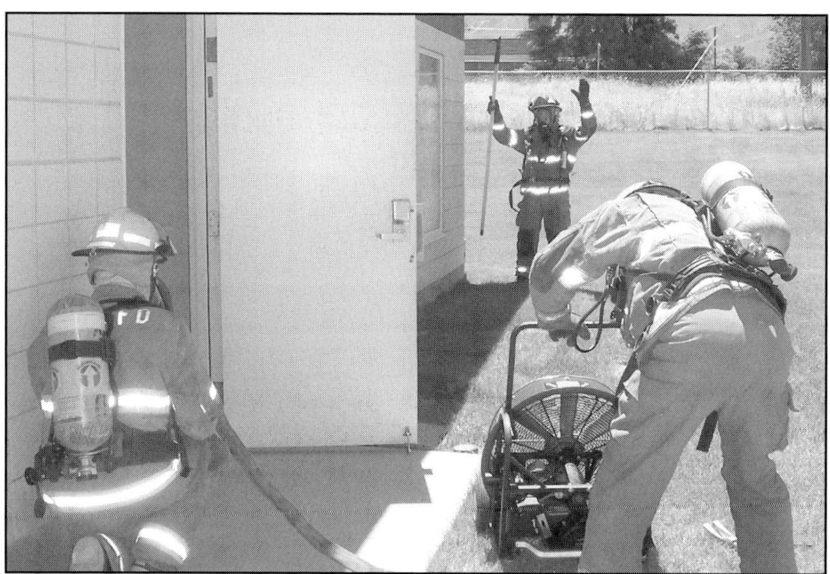

Fig. 5–15 To achieve success, PPA must become the standard for fireground operations on a fire department and not a specialized, rarely used form of ventilation. Courtesy Kriss Garcia

Once firefighters experience the effectiveness of properly done PPA, they simply do not go back to traditional methods. The authors have heard the same story over and over from firefighters on hundreds of departments. They are told that "positive pressure worked so well that we began using it earlier and earlier in the fire, and now we always use PPA." The authors have not heard negative testimony from any of the thousands of students they have instructed who have actually used PPA.

Why PPA Is Dynamic

PPA is truly a dynamic tool for fire attack (fig. 5–16). The word *dynamic* is a versatile one that has different meanings and can be used in many different ways. For instance, something that relates to forces producing motion is considered dynamic. It can mean "forceful" or "energetic." Lastly, something that tends to produce change can also be described as dynamic. All three meanings apply to PPA.

Fig. 5–16 PPA is a dynamic tool for fire attack. Courtesy Richard Moseley

- Dynamic forces produced by the blower act on a lethal atmosphere trapped inside a structure and move it according to laws of physics, giving victims a chance to survive.
- The coordinated ventilation speeds up fire attack and reduces punishment to firefighters, making it dynamic by virtue of being a forceful and energetic element in fire attack.
- PPA is dynamic by virtue of its potential for changing thinking in the fire service and making firefighters look differently at how they attack fires.
- Finally, with PPA, ventilation can be immediately started, stopped, modified, or otherwise altered in response to the changing conditions of a structure fire (fig. 5–17).

Fig. 5–17 This evaporative cooler on the roof became an exhaust opening as pressurization vented products of combustion through the ductwork in the ceiling of the living space. Courtesy Richard Moseley

PPA definitely has dynamic characteristics. It exerts positive influences on the overall fireground operation and results in numerous benefits:

- Heat and smoke are rapidly cleared from the fire structure early in the fire.
- The chances for victim survival are increased.
- Increased visibility assists in search and rescue.
- Clean, cool air replaces the toxic interior atmosphere.
- Crews that would normally be assigned to ventilation are available for search and rescue and fire control.
- Attack lines can be rapidly advanced to the seat of the fire.
- Fire spread is decreased due to pressurization, cooling, and rapid confinement.

Traditional methods introduce many unknowns that are both uncontrollable and unchangeable (fig. 5–18), such as:

- Once completed, vertical ventilation cannot be stopped or reversed.
- Vertical ventilation can introduce fire into a previously uninvolved area if large enough holes cannot be made to let out enough heat to stop horizontal fire extension.
- Horizontally venting by breaking out numerous windows also introduces elements into the fire equation that cannot be altered, modified, or otherwise adjusted to assist interior crews with rescue and fire control.
- Post-knockdown PPV does not provide a benefit for rescue and fire attack and has a risk to firefighters who get between still-active and progressing fire and an exhaust opening.

Fig. 5–18 Unlike PPA, traditional methods introduce many uncontrollable and unchangeable unknowns into the fire equation. Courtesy Richard Moseley

In all the aforementioned situations, victims have a greatly diminished chance of survival after crews put water on the fire in a confined area—again, less than 2% according to the informal survey cited earlier. On a related note, the authors are looking into who makes more live rescues, police officers or firefighters. Early indications are that police officers make more live rescues than firefighters. It appears that police officers who arrive early can enter by keeping below the thermal layer where visibility is relatively good and temperatures are cooler. Firefighters who arrive early and attack the fire without PPA interrupt the thermal balance with water, which dramatically affects visibility and survivability.

On the other hand, PPA works on all types of structures, for most situations where there are victims, and it can be adapted to just about any conditions (fig. 5–19). And if during operations there is any hint of problems or undesired effects, simply turning off the blowers will immediately stop all air movement.

Fig. 5–19 PPA is effective on most types of fires. Courtesy Martha Ellis

An Exciting New Method

Since 1989, the Salt Lake City Fire Department has progressed to having a blower as standard equipment on every engine and truck company (fig. 5–20). Crews are trained to coordinate the blowers with the deployment of attack lines. All companies are trained in, capable of, and expected to provide a coordinated attack on their own. Now an entire crew, or multiple crews, that would ordinarily perform roof ventilation operations can be assigned to other tasks such as forcible entry, search and rescue, extinguishment, and overhaul.

Fig. 5–20 In Salt Lake City, all apparatus carry blowers. This fairly new but well-worn blower that is now out of service is evidence that PPA is regularly used. Courtesy Kriss Garcia

In Salt Lake City, PPA has moved beyond the stage of only being used in one-story structures. It is regularly used in high-rise fires, car fires, dumpster fires, fires below grade, aircraft fires, and just about any other emergency that could benefit from the removal of heat and smoke. Essentially, crews can exhaust any interior environment to the outside through proper placement of blowers in time to benefit fire attack. Strategic and early placement of blowers can alleviate many problems in any interior environment and also can provide benefits when used outdoors, which will be looked at in later chapters (fig. 5–21).

Each firefighter enters a structure protected with equipment worth thousands of dollars that weighs up to 100 pounds or more. Trapped victims are not so protected (figs. 5–22 and 5–23). When faced with opening the nozzle in a toxic, superheated building interior, firefighters should remember this: no one could survive one or two breaths of the toxic, heated atmosphere if the thermal balance is interrupted.

Fig. 5–21 The authors participate in a panel discussion in Great Britain. They have been invited to present their program on PPA to national fire service conferences, state fire training academies, individual fire departments across the country, and to firefighters in foreign countries. Courtesy Kriss Garcia

Fig. 5–22 Trapped victims awaiting rescue may be protected only by bed clothes. Courtesy Kriss Garcia

Fig. 5–23 Each firefighter enters a structure protected by equipment worth thousands of dollars that weighs up to 100 pounds or more. Courtesy Kriss Garcia

As with any fireground operation, there are precautions to consider when using PPA. These will be examined in the next chapter.

Beyond the Basics

By now the reader should have enough background to begin training with basic PPA and PPV. This section wraps up all the loose ends. It explores other uses for blowers and goes into more detail on aspects that have already been discussed.

This section also looks at some of the cultural and organizational aspects of introducing PPA or any new method in the fire service. This discusses how to approach training, offers some examples of detailed evolutions, and explains how to win over the skeptics. It finishes off with a review of 20 of the most common questions asked about positive pressure methods.

6

Precautions for Using PPV and PPA

PPV and PPA contribute greatly to the safety and effectiveness of the services firefighters provide. The vast majority of the time, crews will successfully use positive pressure by simply placing a blower at the back of the initial attack crew and ensuring an exhaust point has been established. However, no matter how simple it may appear, there is far more to it than just that. Because pressurization occurs during the very dynamic initial stages of fire attack when free burning is underway, PPA does have the potential to contribute to or cause some problems. The good news is that all these problems can be dealt with by adequate training and procedures (fig. 6–1).

Everything considered, PPA is by far safer for firefighters and more beneficial for victims than putting crews above a fire on a roof or beginning pressurization after the fire is knocked down. It is also a far more effective means of making rescues and protecting property.

This chapter explores in greater depth some of the situations referred to in earlier chapters where caution and careful application of positive pressures are required. It is not meant to be a detailed lesson of related conditions such as backdraft, but rather a general discussion of observed and theoretical conditions relating to PPA and how to address them properly and safely.

Fig. 6–1 The authors train with firefighters in British Columbia. Training is essential to have safe and effective PPA. Courtesy Kriss Garcia

The Importance of Training

Firefighters rolled up in front of a 50-year-old, two-story, ordinary constructed, 1,500-ft^2 home. The fire was located in the kitchen area on the main floor. First-arriving crews quickly set up a blower at the main entrance and began advancing an attack line through the front door. Meanwhile, the home's resident was attempting to exit from an open window in the rear of the building upstairs. As the airstream from the blower was directed inside and the attack crew moved in to hunt down the seat of the fire, heat and smoke belched out the open window in which the resident was waiting. The resident was forced to make a hasty detour with the help of an air conditioning unit that, luckily, was located adjacent to the window.

The resident in this real incident was a victim of improperly done PPA. As the blower ventilated the interior and cleared out the smoke and heat, the attack crews were able to quickly advance to the fire. However, smoke and heat were pushed out through the only exhaust opening at the time—the window opened by the resident. Attack crews did not wait for a report on the status of the exhaust opening before beginning ventilation, and it is possible that nobody checked for one at all. If not for the nearby air conditioner, the resident could have been seriously burned or may have been forced to jump.

Luckily, this type of situation does not occur often. PPA has the potential to be hazardous when proper care is not taken, or if it is used without adequate training. Although (or, perhaps, because) situations like this happen very infrequently, training and practice cannot be overemphasized. Attempting to use PPA with inadequate training increases the possibility that pressurization of a fire could result in unnecessary injury or other detrimental impacts (fig. 6–2).

Fig. 6–2 PPA has the potential to be hazardous when proper care is not taken. Training and practice cannot be overemphasized. Courtesy Ray Schelble

Like any tactical operation on the fireground, having the skill and capability to perform an evolution is only part of the battle. Another important part is recognizing its limitations. At a minimum, everyone on a department, from the firefighters who position the blowers to the IC who directs the operation, should know and understand the basic procedures for and the limitations of pressurization.

Paying attention to safety considerations is essential when using PPA, but not to any greater extent than with any other fireground operation. In fact, as has already been discussed, PPA is far safer for firefighters and victims. Situations where PPA is not advisable are few and far between.

Proper training in PPA consists of understanding the basic principles that govern pressurization, learning proper fireground procedures, knowing when to use PPA, and continued schooling and practice. Departments that do this on a continuing basis will be able to use PPA and PPV as valuable tools to consistently benefit their department and community.

Exhaust Point Safety

Using a high-volume blower when active burning is under way in a structure creates great benefits in an otherwise hazardous and punishing operation. PPA will not "fan the flames" and intensify a fire. It can, however, create a potential hazard zone for firefighters, victims, and exposures near the exhaust opening. It also can create a hazardous situation for firefighters who may be working in the area between the main body of fire and the primary exhaust opening.

Belching flame and smoke

One of the things about PPA that people are least prepared for is what happens at the exhaust opening (fig. 6–3). When a blower increases the pressure in a building and forces fire, heat, and smoke out of an exhaust opening, startling things happen (fig. 6–4). Smoke billowing from the opening gives the appearance of a raging inferno, which in reality is just the heat and products of combustion being forced from the rapidly clearing interior environment. At worst, hot gases will ignite as they come in contact with the environment outside the opening. In reality, the heated products of combustion from inside are able to mix with outside air and reach the optimum pressure and oxygen requirements for ignition. The display can create anxious moments for people who are not prepared for it. It may also have something to do with the impression that pressurization "fans the flames," even though this is not the case at all.

Fig. 6–3 Before pressurization. One of the things people are least prepared for is what happens at the exhaust opening. Before starting pressurization, smoke rises from the exhaust opening. Courtesy Kriss Garcia

Fig. 6–4 After pressurization. Fire, heat, and smoke billowing from the opening give the appearance of a raging inferno. In reality, this is just the heat and products of combustion that are being cleared from the interior. Courtesy Kriss Garcia

Because of the nature of the exiting heat and other products of combustion, the area near the exhaust opening must be considered a potential hazard. Although not common, it could have the potential to injure someone if safety is overlooked. Firefighters must always keep in mind that any broken window or other opening on the outside of the building has the potential to become an unintended exhaust opening if it is not closed off from the airflow of the blower. In most cases, this hazard diminishes quickly as the heat and products of combustion from the interior are replaced by cooler air from outside.

If any problems develop after beginning pressurization, the blowers can be turned down, repositioned, turned off, or otherwise adapted to ensure that the ventilation meets the needs of the situation. Conditions should be constantly reevaluated, and tactics should be adapted to meet needs as they arise.

Protecting firefighters

Firefighters must take care to keep crews and the public away from the exhaust opening and to not use it for access. A firefighter breaking a window or opening any exhaust opening must be careful to stay out of the way of the rush of flame and heat that may forcefully vent (fig. 6–5). Firefighters must always work from a position to the side that affords quick escape following completion of the exhaust opening.

There may be times when, for some reason, firefighters will access the interior through a window or door that could become the main exhaust opening. A tactic used by some departments is called vent-enter-search (VES). In VES, firefighters enter a building from exterior openings, usually making extensive use of ladders. Firefighters should not attempt PPA if VES is being used, or in any case when firefighters are in locations that could be in the path of exhausted products of combustion. The best way to avoid these types of exhaust point problems is to enter only from the ventilation opening when using PPA.

Pressurization should never be started until all personnel are safely out of the potential exhaust stream or any other opening that could become an exhaust opening. As mentioned earlier, pressurization immediately starts to improve conditions at the floor level, where many victims will be located. However, care must be taken to protect firefighters who may be standing up or working above the floor level in an area that will become the path of travel for the exhaust. This is more of a problem with post-knockdown

PPV than with PPA. Again, establishing PPA procedures that require the initial blower to go at the main access point at the back of the initial attack crews generally eliminates this hazard. Operating in this manner is far safer for firefighters than beginning PPV post-knockdown, not to mention the benefits it provides to victims.

Fig. 6–5 When creating an exhaust opening, firefighters should wear full protective gear, keep off to the side of the opening, and work from a position that affords quick escape following completion of the opening. Courtesy Kriss Garcia

Although the exhaust point unquestionably has the potential to create a serious problem with PPA, the fire behavior and the hazards associated with it are very predictable and easily dealt with. Still, PPA is sometimes criticized for this even though the precautions are not any more serious than precautions observed with vertical ventilation or post-knockdown PPV. Would firefighters be assigned inside an attic before the roof is opened for conventional vertical ventilation? They absolutely would not! Neither should firefighters be forced to work near the exhaust opening.

Victims at the exhaust point

The first priority in the fire service is life safety, and PPA can play an important role in saving victims who otherwise would not survive. It is the firefighters' responsibility to, as much as possible, ensure victims are not located in areas that may be hazardous to them (fig. 6–6).

Fig. 6–6 In this test fire, a rooftop crew still works to create a vertical ventilation hole while PPA clears the interior. Before pressurizing the building, crews must be certain a victim is not awaiting rescue at an exhaust opening. Courtesy Ray Schelble

The example at the beginning of this chapter shows what can happen when a victim is standing in an open window that becomes the exhaust point. It is critical that the crew member assigned to the exhaust opening report to the IC before the building is pressurized. Again, most of the time it is not an issue, as the fire has self-vented by the time crews arrive, and no one could survive for long in this area without PPA.

Pressurization should never be started when a victim is at a window awaiting rescue. This could have grave consequences. First priority must be given to known victims, and their safety must be a primary tactical consideration. After victims at potential exhaust openings are rescued, tactics can be adjusted to accommodate further search and rescue operations.

Sometimes victims cannot readily be removed to safety because of access or other problems. If they can communicate with firefighters, they could be instructed to shut the door to the room they are in while waiting for a ladder rescue to be made. Closing the door to the room isolates the area, which means the open window will no longer be an exhaust point of any consequence. The building can be pressurized to help crews gain access through the interior. In this type of situation, the firefighter making contact should not leave the victims and should communicate their location and condition to command. If deteriorating conditions appear to be putting the victims in more danger, command should be notified and tactics adjusted. At all costs firefighters should not allow an opening where a victim is awaiting rescue to become the primary exhaust opening.

Exposure hazards

Any buildings or objects within 6 ft of the exhaust opening should be considered exposures that need to be protected, moved, or at least examined for fire extension. Without exception, all personnel and spectators must be kept a safe distance away (fig. 6–7). Experience has demonstrated that a combustible building within 6 ft of the exhaust opening could be ignited by the heat and flame exiting through the exhaust opening. In many of the authors' test fires in acquired buildings and at actual fires, they have seen flame vent up to 4 ft horizontally from the primary exhaust opening. As with any exposure, playing a hose stream on the exposed structure or object will provide adequate protection.

Fig. 6–7 A fire in the front of a residence vents after a blower is directed into the back door. Any exposures within 6 ft of an exhaust opening should be protected, moved, or at least examined for fire extension. Without exception, all personnel and spectators must be kept a safe distance away. Courtesy Alex Groves Photography

The same exposure hazard exists if the exhaust opening is beneath a horizontal overhang, such as an upper-story porch or a roof soffit area. The possibility of fire extension must be considered and adequate measures taken to protect these areas. Firefighters must make sure the venting fire is not directed into an uninvolved attic area, especially when attic vents are close by.

During test burns in acquired buildings, fire venting from the exhaust opening impinged on the roof soffit far more before the pressurization began than after. Pressurization actually forced the fire horizontally beyond the overhang. The bottom line is that any exposure must be identified, evaluated, and protected regarding fire extension anyway, whether PPA is used or not.

There is usually leeway in where to locate an exhaust opening if the fire has not already self-vented. Firefighters must consider exposures, overhangs, and a clear escape route when making the decision.

Make Sure the Fire Is Out

Rekindles, fires that "miraculously" spring back to life after fire crews supposedly have them completely out, are the bane of the fire service. They may seem to occur miraculously, but they do not. As all firefighters well know, a rekindle is viewed as a failure to do one's job thoroughly and competently. Any argument that has a chance of diffusing the blame for a rekindle is often pursued.

Some rekindles that occurred during the early years PPV was being used were blamed on the blower supposedly forcing fire through intact areas into hidden areas or fanning flames in hidden void spaces. PPV was labeled a method that could keep a fire small and hidden and then magically fan it back to life after the firefighters left, a myth that continues to hinder acceptance of PPA. The reality is, rekindles happened long before anyone ever thought of using positive pressure. Without proper overhaul, rekindles can happen in any difficult fire situation, whether positive pressure methods are used or not (fig. 6–8).

Fig. 6–8 Rekindles, or fires that "miraculously" come back to life after firefighters leave, are the bane of the fire service, and some have been blamed on the use of positive pressure. Crews at this fire did what was necessary to prevent a rekindle whether positive pressure is used or not. They opened up walls to root out hidden fire. Courtesy SLCFD

Blowers and rekindles

The authors have closely examined numerous incidents where rekindles were blamed on the use of a blower, including some that occurred while crews were still on the scene contemplating additional overhaul. Based on what they found, they have resolved that the problem lies not with PPA but rather with how it was misused or how firefighters did not recognize the different environment it creates.

Seasoned fire crews use their senses, honed over decades, to determine when a fire is completely extinguished and overhaul can stop. Their decision is based primarily on the decreased heat levels and the absence of drift smoke. When things cool down and there is no more smoke, the fire is considered out. However, it may not be.

When using positive pressure to ventilate, crews evaluating the need for additional overhaul are in an environment that is altered by the blowers. They must take this influence into account. The cooling and air movement produced by today's high-volume blowers have a tremendous impact on the interior environment of a building. Blowers easily overcome otherwise-obvious drift smoke before it has a chance to linger. Blowers also remove the transient heat that would be detectable in still air. With a blower running, crews depending on their past experience with their senses to detect fire extension may believe the fire is completely out, when in reality, further overhaul and examination are warranted (fig. 6–9).

Fig. 6–9 Blowers alter the interior environment at a fire and therefore reduce or eliminate drift smoke and other indicators of hidden fire. With a blower running, the fire may appear to be out when it is not. Courtesy TCVFD

Turn off the blower

The only way to eliminate rekindles is with complete overhaul. The only way to make sure the area in question needs no further overhaul is for crews to turn off the blower before they make their final assessment. When determining whether additional overhaul is necessary, fire crews should turn off the blowers for 10 to 15 minutes and then reexamine the area. If the fire is not completely out, it will make it far easier to tell if additional overhaul is necessary.

When in doubt, firefighters should open it up (fig. 6–10)! They should follow the advice to "open early and open often" when it comes to overhaul. If there is any question, they should open up any areas that have a chance of containing hidden fire to visually check the area. If fire has impinged on a space, the dollar loss will be the same, since it will have to be repaired and rebuilt anyway.

In any fire situation where there is a possibility for a rekindle, a fire watch should be posted or detailed instructions left with a responsible person who remains on the scene afterward.

The use of a Class A foam additive decreases the surface tension of water and allows more complete penetration into combustible material. This has considerable value in decreasing the chance of rekindle from incomplete extinguishment. Again, firefighters should turn off the blower for 10 to 15 minutes and reevaluate, and then apply foam to the area of concern.

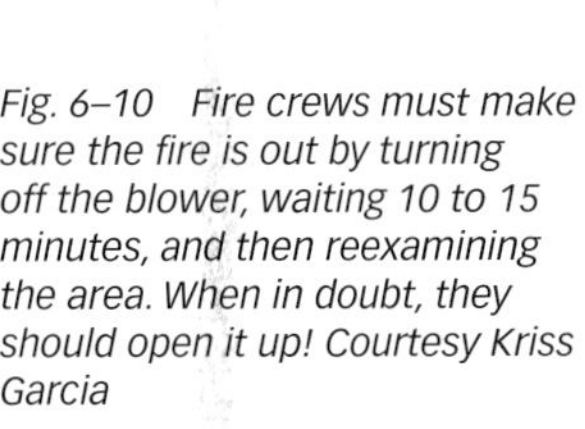

Fig. 6–10 Fire crews must make sure the fire is out by turning off the blower, waiting 10 to 15 minutes, and then reexamining the area. When in doubt, they should open it up! Courtesy Kriss Garcia

Problem Situations

Positive pressure has wide-ranging applications in the fire service, but it is by no means a method to be used for every situation. PPA can aggravate problems with backdraft conditions, combustible dust environments, and when flammable vapors are present. Carbon monoxide alarms and weather can cause additional problems if pressurization is used without a complete evaluation of the entire scene.

Backdraft conditions

Backdraft is a serious threat to firefighter safety. Backdraft conditions develop when a fire has burned in a closed building or an enclosed area or a void inside a building for a long enough to consume the available oxygen. The fire has filled the space with plenty of preheated, flammable products of combustion. When the oxygen level drops far enough, free burning stops, and the fire enters a smoldering stage. The building is charged with hot, flammable gases and high levels of heat. Introducing fresh air to a fire with backdraft potential will rejuvenate the fire quickly, possibly with explosive and potentially deadly force (fig. 6–11).

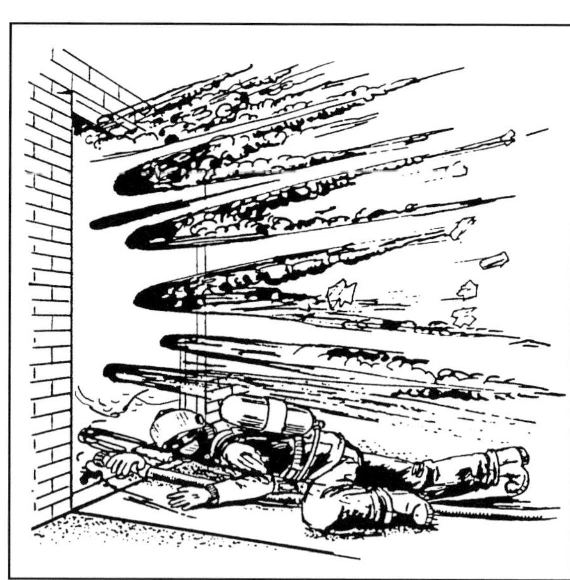

Fig. 6–11 Backdraft is a serious threat to firefighter safety. Introducing fresh air to a fire with backdraft potential can have explosive results. Courtesy SLCFD

The authors' conclusions regarding the use of positive pressure when a backdraft condition exists are based on theoretical analysis rather than on actual experience using positive pressure methods on imminent backdraft conditions. Producing conditions that will lead to backdraft in a training scenario is difficult. Often what people call a backdraft is actually a flashover. Most firefighters work their entire career without witnessing a backdraft.

Identifying backdraft potential. As a general rule, if flame is visible, the backdraft potential is very low. To identify a fire with the potential for backdraft, firefighters should look for the following indicators in an enclosed area:

- Thick smoke that may vary in color from black to dark yellow to brownish black. Burning hydrocarbons, plastics, or similar sources may produce similar-colored smoke that does not indicate backdraft potential.
- Smoke forcing under pressure or puffing from gaps in doors, windows, or other openings.
- Air movement inward through openings, especially a rapid air movement when entry is made to the building.
- Lack of active flame.
- Windows that are unbroken and discolored on top from smoke, or that are covered with oily or watery stains from smoke condensation.
- Windows, doors, and walls that are unusually hot to the touch.
- Even when backdraft conditions are not present on arrival, they can develop during firefighting operations. Sounds that seem unusually muffled when inside a building indicate the possibility of developing backdraft conditions. Because PPA provides ventilation early in the incident, it should almost certainly eliminate this type of backdraft occurrence except, perhaps, in a void space (fig. 6–12).
- Backdraft conditions can also develop in voids or other closed-off areas within a structure (fig. 6–13). This is a common backdraft condition, but aggressive overhaul will make it less likely to happen.

Fig. 6–12 The many openings in the building used for this training fire give it an extremely low potential for backdraft. Courtesy Rowe Harrison

Fig. 6–13 Even when conditions may not promote the development of backdraft conditions in an entire building, they can still occur in voids or enclosed areas. This is a common backdraft scenario, but aggressive overhaul will make it less likely to happen. Courtesy Richard Moseley

Assessing backdraft options. When true backdraft conditions are present in a structure, temperatures have reached more than 1,000°F and have been so for a considerable time. The prospects of saving life in this environment are zero. The interior atmosphere is highly lethal, which ensures there is not enough oxygen in the area for anyone to survive. If enough oxygen were available for victim survival in the fire area, the fire would still have enough to continue to free burn. This being the case, no rescue operation is necessary, since anybody exposed to those high temperatures and the toxic, oxygen-deprived environment could not possibly survive (fig. 6–14).

Fig. 6–14 When backdraft conditions are present in a building, the high interior temperatures, low oxygen levels, and lethal atmosphere ensure that chances for a rescue are zero. Everything in the interior will have already sustained major damage. Courtesy SLCFD

From a property conservation standpoint, under these conditions items in the structure will have received extensive damage from smoke and heat. The value for most items will be nonexistent. Even the building itself may have seriously diminished value given the circumstances and conditions.

Everything considered, a situation that has no chance of supporting life and can provide little or no benefit for property conservation should dictate the course of action. The conventional approach is to first ventilate high to release superheated flammable smoke from the building, and then create low openings for cooler air to enter. This is still the course of action that should be considered. However, when considering sending a crew to the roof for vertical ventilation, there is something important to keep in mind. Except for perhaps a fire resistive building, the high temperatures associated with the backdraft conditions could have weakened the structural integrity of the roof structure. Interior firefighting operations should not begin until free burning is once again taking place or obvious fire behavior conditions indicate that backdraft conditions no longer exist.

Firefighters operating around a structure under these conditions must do so from safe positions. Until ventilation is complete, crews should keep clear of doors or windows that could blow out as the interior atmosphere ignites and expands with explosive force (fig. 6–15).

Fig. 6–15 Conditions that indicate a potential for backdraft, such as a fire burning for some time in a closed-up space, require a careful approach when it comes to ventilation. Courtesy Gina Bell

In this situation, any rooftop operations should be carefully considered by weighing the risks from possible structural weakness of the roof or from explosion against the benefits. An IC would have a hard time justifying the logic of deploying firefighters on top of what is in reality an assumed imminent explosion. Instead, the authors advocate exhausting the explosive atmosphere from safe positions and into an area that will not cause direct exposure extension. This process should begin only after master streams and other resources are in place that can confine the fire and protect any threatened exposure.

After the initial venting operation, blowers should be properly placed for PPV and allowed to operate for several minutes before firefighters enter the building. Firefighters should not enter a building that is exhibiting backdraft conditions until the IC is confident that the interior has been cooled with pressurization combined with water application from the exterior. Only after the building is no longer considered to be a backdraft threat should firefighters be allowed to enter. The process of decreasing the temperatures fast enough to prevent the products of combustion and the building from igniting is one where knowledge and experience are valuable. A seasoned IC is the key to ensuring the safe deployment of resources.

Combustible dust and flammable vapors

Pressurization can be extremely hazardous in a potentially explosive atmosphere. Moving an accumulation of combustible dust or flammable vapor into another area that may pose an ignition threat is extremely dangerous. Any air movement runs the risk of creating turbulence that could worsen a bad situation. Pressurization should not be used until the IC has a complete understanding of all conditions that could affect the operation.

Dust hazards. After life safety considerations, the first priority in an incident involving potentially explosive dust is to make the environment safe by applying water gently, as a mist if possible. A water mist mixing with the finely divided particles reduces the potential that they can become airborne, thus decreasing the explosiveness.

In situations such as a cereal or grain silo where air turbulence may move dust into an area that has a possible source of ignition, positive pressure absolutely should not be used (fig. 6–16).

Fig. 6–16 PPA absolutely should not be used in a structure such as a grain or cereal silo, where air turbulence could move dust particles to a possible ignition source. Courtesy Kriss Garcia

Vapor hazards. When it becomes necessary to ventilate an area containing flammable gases with a low flash point or low ignition temperature, such as vapors from a flammable liquid spill, utmost caution is required (fig. 6–17). Consulting with hazardous materials technicians before taking any action is highly recommended. Any potential paths vapors may travel after being exhausted from a structure must be considered, as well as ensuring the vapors cannot contact any possible ignition source once they are exhausted. Ventilating with an internal combustion engine, which is also a possible ignition source, should be evaluated on a risk/benefit basis by the IC. Safe, explosion-proof, high-volume electric blowers are available for these types of operations.

When dealing with flammable vapors, firefighters should not take action unless they are sure that what they are doing will not make the situation worse or create a new emergency (fig. 6–18).

Fig. 6–17 Extreme care must be taken when using blowers whenever explosive vapors may be present. A natural gas explosion leveled this Salt Lake County home before firefighters arrived. Courtesy Richard Moseley

Fig. 6–18 Explosion-proof, high-volume blowers are available, but firefighters should not take action unless they are sure that what they are doing will not make the situation worse. Courtesy Kriss Garcia

Smoke explosion potential

Conditions for a smoke explosion may look similar to backdraft conditions, but there are some important differences. The limiting factor in a smoke explosion is an ignition source. The limiting factor in backdraft is oxygen.

Smoke explosion conditions form when fire uses up the oxygen in a space to the point where the fire burns itself out. This leaves the space charged with products of combustion, but because of the nature of the products being burned, the smoke itself is flammable. Often, this condition creates heavy amounts of thick, acrid black smoke, which may only be warm or even could be cold. It often occurs because of combustion involving formaldehyde-based insulation or other synthetic materials.

This condition is relatively rare and may be difficult to distinguish from a potential backdraft. It should be handled similarly to a combustible dust situation. If using pressurization on cold smoke, firefighters should not exhaust it into an area that has an ignition source. Also, an explosion could occur if fresh air allows the original fire to reignite. Firefighters must use extreme caution when smoke explosion conditions are suspected.

Carbon monoxide alarms

The first priority on a call for a CO alarm is, as always, life safety. First consideration should be for rescuing and treating victims who are having negative reactions to a high-level CO environment. If this is not the case, which is what firefighters commonly find, the next priority is usually to ventilate the environment and replace it with fresh air (after first making sure that the CO alarm is not malfunctioning, of course).

In an atmosphere in which the sole contaminant is CO, ventilating with an internal combustion blower usually will not adequately reduce already low levels of CO (fig. 6–19). After the threat to life safety has been addressed by determining the levels of CO through metering, finding the possible source should be the next tactical priority.

Until the suspected area of contamination is determined, trying to remove this odorless, tasteless, invisible gas by using a tool that creates CO only serves to complicate the process of determining the source. Any internal combustion engine produces CO as a by-product. Gasoline-powered blowers built for the fire service have exhaust extensions available that are supposed to reduce CO introduced into a building. However, in the authors' opinion, these extensions are cumbersome and only reduce CO in a building by 5%.

Blowers manufactured after approximately 1995 that have overhead valves produce considerably less CO than blowers made earlier. Use of natural breezes and/or electric, air, or water-driven blowers or building ventilation systems will be the best choice to remove CO.

Fig. 6–19 Electric blowers or other means can be used for ventilation on a call for a CO alarm. Gas-powered blowers normally will not adequately reduce nuisance levels of CO. Courtesy Kriss Garcia

Adjusting for weather

Wind can be an ally or an enemy when using PPA. When at all possible, the attack should not be made from the leeward side of the fire, just as it would be without pressurization.

Locating the ventilation opening and attack entrance on the windward side allows nature to assist with interior pressurization. It provides a safe working environment for crews while ensuring they are not caught in a crosswind of competing ventilation currents.

In a steady wind, simply opening windows and doors will create a horizontal air flow. In this type of operation, good communication is necessary to keep the IC or tactical supervisors apprised of the actions taken. Care must be exercised to keep firefighters and victims away from any exhaust openings on the leeward side of the building, the same as if there was no wind and they were using PPA.

The opposite occurs if windows on the windward side of the dwelling are opened, and the fire is attacked from the leeward side. This approach is similar to placing blowers at the front and back doors. There is essentially no exhaust opening, which forces crews to work in an environment that is unstable and unpredictable.

In test burns and actual fire incidents, the amount of consistent wind it takes to become a problem varies depending on the size of the windward opening. As a general guideline, in any consistent wind of 10 mph or more, or gusty winds of 30 mph, the attack should be adapted. The initial crews will need to enter the windward side of the building. Of course, the exhaust opening should be on the lee side of the structure.

As storm fronts come and go, the changes in barometric pressure seem to have little effect on pressurization of a structure (fig. 6–20). The air pressure in the building is equalized with the exterior pressure, regardless of barometric pressure. The blowers simply raise that pressure an additional 0.1 psi to 0.2 psi relative to the pressure outside.

The next section delves into more specific uses for positive pressure, training, and implementation, along with some of the more popular questions that arise.

Fig. 6–20 As weather fronts come and go, the changes in barometric pressure seem to have little effect on pressurization of a structure. Courtesy Ray Schelble

7

PPA and Beyond

Up to this point, positive pressure methods have been discussed strictly from the perspective of ventilating buildings and as a component of coordinated interior fire attack. These types of opportunities alone justify a department's investment in training and the purchase of blowers, and will typically make up most of the workload of the blowers.

But there is more. The blowers used for positive pressure can also be used for a variety of situations in addition to what has already been discussed. In this chapter, the role of sequential ventilation in clearing compartmentalized buildings gets a further look, as well as some of the challenges in ventilating high-rise buildings.

Effective Uses for Positive Pressure and Blowers

Blowers create a positive pressure differential in an enclosed space, but they can also be very effective when used outside. Just about any time the environment compromises victim survival or rescuer effectiveness and safety, blowers and pressurization should be considered to create more favorable conditions (fig. 7–1). Their ability to move large volumes of air can help protect firefighters and victims in a wide variety of situations, including vehicle fires, dumpster fires, aircraft fires, chimney fires, and confined space rescue. For incidents that require moving a large volume of air, a mobile ventilation unit can help.

Fig. 7–1 Salt Lake City firefighters use two blowers (visible in front of firefighters working in background) to keep fire in an overturned semitrailer away from a vehicle pinned underneath that has a woman trapped inside. Pressurization should be used just about any time the environment compromises victim survival or rescuer effectiveness and safety. Courtesy Martha Ellis

Vehicle fires

A burning vehicle creates a particularly unhealthy environment. Vehicles contain an abundance of various plastics and other man-made materials, as well as dangerous quantities of flammable liquids. In addition, metals used in their construction can intensify a fire and make it extremely difficult to extinguish.

Before attacking a vehicle fire, firefighters should set up a blower on the upwind side about 10 ft to 12 ft away and direct the airstream at the vehicle (fig. 7–2). Crews then attack the fire with the blower to their backs.

Although technically this is not PPA, the airstream helps force toxic products of combustion away from firefighters because of the positive pressure differential created at their location. This can greatly improve visibility and is especially valuable during overhaul (fig. 7–3). As with structure fires, after overhaul, firefighters should turn off the blower for 10 to 15 minutes and inspect the vehicle thoroughly to make sure the fire is completely out.

Fig. 7–2 Before attacking a car fire, firefighters should set up a blower on the upwind side about 10 ft to 12 ft away and direct the airstream at the vehicle. Courtesy Tempest Technology, Inc.

Fig. 7–3 A blower improves visibility and safety in this Salt Lake City Fire Department training exercise. Courtesy Ray Schelble

Besides protecting firefighters, utilizing blowers will help to quickly determine if there are any victims inside. Vehicles rank third as areas where victims are likely to be found.

Dumpster fires

Pretty much the same thing goes for dumpster fires as for vehicles. There is no telling what people will throw away. The list of hazardous materials that can be found in these receptacles may be as long as the list of hazardous materials in existence.

Generally, the blower should be placed upwind and the attack made with the blower to the firefighters' backs. Tilting a blower upward will usually improve effectiveness. When a building is nearby, the blower must be positioned to keep heat and products of combustions from impinging on it (fig. 7–4).

Fig. 7–4 For a dumpster fire, firefighters should set up the blower to the firefighters' backs. To protect an exposed building with fire impinging, they should set up the blower alongside the building with the airstream directed parallel to the wall, between the wall and the dumpster. Courtesy Tempest Technology, Inc.

Aircraft fires

Aircraft are built of lightweight materials and concentrate large numbers of people in close quarters. Any aircraft fire is something to be taken very seriously.

In general, the same approach for positive pressure that applies to structure fires should be followed with aircraft, but a caution is in order here. Although PPA has been well documented in structure fire situations, little has been done to study it in aircraft fires. Logically, everything should behave the same in an aircraft cabin fire as in a structure fire. The authors have conducted training on aircraft with theatrical smoke that had very favorable results, but to their knowledge it has not been used in training with actively burning fire. At this writing, they have been able to document only one case in which positive pressure was used on an actual fire. In other words, PPV has value in aircraft fires that the authors have documented, but PPA should be attempted only with great caution.

In an incident involving fire or smoke in an aircraft parked at the gate, how to proceed depends on several factors. If protecting the terminal from fire is the primary concern and there is nobody on board, backing the Jetway off the aircraft may be the logical solution. In some situations, passengers may have to exit through the Jetway, or firefighter access from the terminal may be required or desirable (fig. 7–5). In these cases, a blower or blowers can be positioned inside the terminal at the entrance to the Jetway. As long as the boot is against the aircraft, the passenger compartment will be pressurized almost as if the blower were positioned right at the entrance of the aircraft. Open access doors in the Jetway can cause a major reduction in pressurization, but a blower should still be able to maintain effective pressurization with the normal gaps between the boot and the aircraft.

Several basic cautions for positive pressure are particularly important to aircraft emergencies. First, under no circumstances should firefighters begin positive pressure when an opening that is being used for access or egress could become an exhaust opening. Second, pressurization should never be used when flammable liquids are actively burning. The air moved by the blowers causes unpredictable movement of vapors, which creates serious hazards. Third, to safely use positive pressure in an aircraft, there must be an exhaust opening.

Fig. 7–5 Blowers in the concourse were used to pressurize this aircraft at the gate. Passengers were evacuated through the Jetway. Courtesy Salt Lake City International Airport

Chimney fires

A fire in a chimney can be more effectively extinguished with a pressurized attack than any other way. Pressurizing the structure and restricting exhaust openings ensure that fire, products of combustion, and burning material will be confined to the fireplace.

Most chimney fires are handled by the application of a small amount of water or dry chemical to the firebox, which allows steam or powdered chemical to extinguish the fire in the chimney. For the most effective ventilation, firefighters should pressurize just the floor level that the fireplace is on. This ensures that as the firebox is opened and extinguishing agent is introduced, most by-products will be forced up the chimney. Any windows or doors adjacent to the fireplace opening should be kept closed so maximum air volume flows up the chimney.

With positive pressure operating, the authors have had greater success applying small amounts of water from a pressurized extinguisher into the firebox rather than dry chemical or CO_2 extinguishers. Water used sparingly allows gradual cooling of the firebox while steam is forced into the chimney area. If the fire should spread from the chimney into the attic area, steam from the efforts below will assist in extinguishing fire in those areas also.

When openings have to be made into the attic space to check for extension, access through the ceiling should be made first. When there is an attic cubbyhole access, crews can generally check the space, causing little, if any, property damage. If no attic access can be readily found, creating an access hole in the ceiling adjacent to the chimney is often less expensive for the homeowner to repair.

Opening a roof to check for attic fire extension should be avoided. Because of the effects of weathering on roof material, matching the roof covering is often not possible. The entire roof area may have to be replaced when a simple, easily repaired hole in a drywall ceiling would have provided equally effective access.

Firefighters should remember that whenever possible, fires should be attacked from the level they are on or from below. Only as a last resort should attack be made from above. Crews are far more at risk and much less effective when operating above a fire.

Confined space rescue

Many incidents involving trapped victims in confined spaces can be helped with the judicious use of positive pressure (fig. 7–6). A toxic environment can quickly be replaced with fresh air to increase the chances for victim survival and make it easier for rescuers to do their jobs. When using positive pressure for confined space rescue, firefighters should keep in mind the following points:

- Positive pressure should not be used in unlined dirt tunnels or trenches. The air movement dries out the dirt, which could make collapse more likely and can stir up contaminants and dust (fig. 7–7).

- Pressurization never eliminates the need for rescuers to protect themselves with SCBA and appropriate protective gear. Also, firefighters should take precautions to protect the victims.

- Monitoring the interior area for toxic or flammable gases must be done when working in any confined space location (fig. 7–8).

- High-volume electric blowers do not add contaminants and should be considered the preferred option when ventilating a confined space environment. As discussed in chapter 6, even with exhaust extensions, gasoline-powered blowers still introduce CO.

- Firefighters should not use gasoline-powered blowers or electric blowers that are not rated for use in explosive environments to ventilate a potentially flammable or explosive atmosphere.

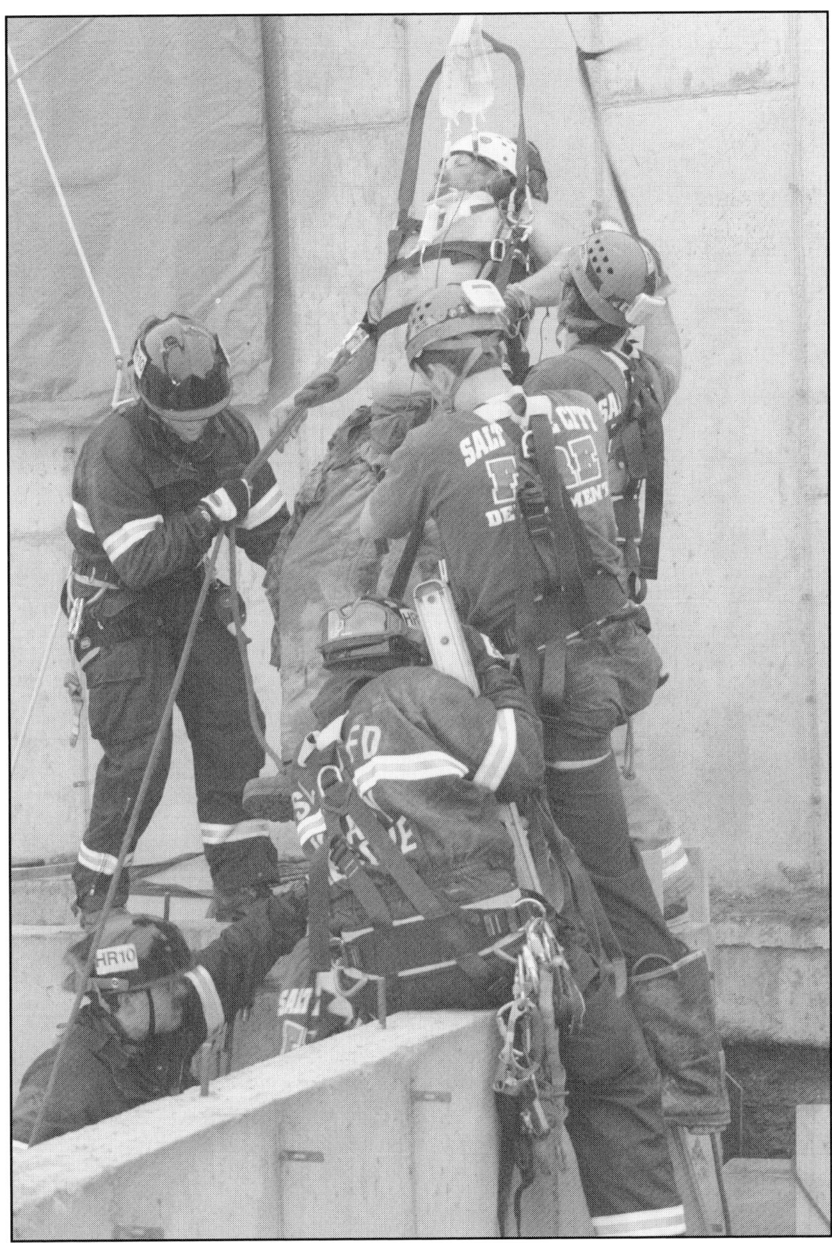

Fig. 7–6 Incidents involving trapped victims in a confined space can be helped with the judicious use of positive pressure. Courtesy Martha Ellis

Fig. 7–7 Rescuers and victims alike should take precautions to protect themselves in confined space situations. Firefighters should not use positive pressure in unlined dirt tunnels or trenches. Courtesy Tempest Technology, Inc.

Fig. 7–8 Among other precautions, confined space rescuers must still wear full protective gear and monitor for lethal or explosive gases. Courtesy Tempest Technology, Inc.

Sequential ventilation

Most buildings are compartmentalized to some extent by rooms, offices, apartments, floors, and the like. Common sense says that smaller areas ventilate faster than larger ones. Crews using positive pressure can take advantage of compartmentalization to more effectively and efficiently ventilate large structures. Opening and closing doors or windows can divide a compartmentalized building into smaller areas that, when ventilated in a systematic sequence, can quickly clear contaminants, often without moving the blower.

Initially, firefighters should close off all but one area to pressurization and open an exhaust opening in the pressurized area. When the first area is cleared, they should close off the path of airflow into it and open up a ventilation and exhaust point in the next area. This procedure should be repeated until all areas are cleared.

For instance, a floor in an apartment building or office building can be cleared sequentially by first pressurizing the hallway. Then a door is opened and an exhaust opening (usually a window) is created in one apartment at a time, closing each off when it clears (figs. 7–9 and 7–10). Clearing each apartment one by one down the hallway will quickly and effectively remove all the contaminants. A floor with offices can be cleared in the same way.

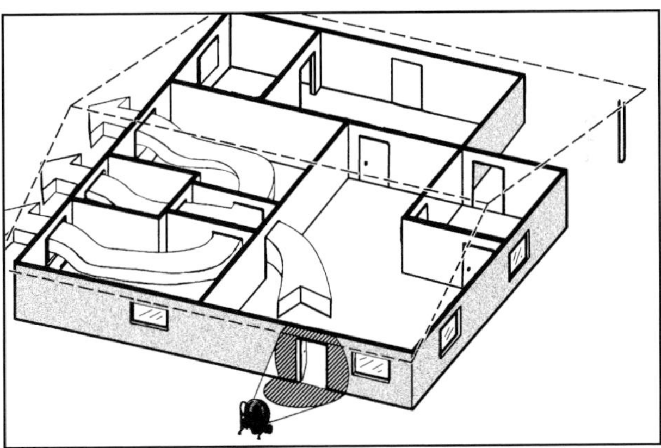

Fig. 7–9 Figures 7–9, 7–10, and 7–11 show the general procedures for sequential ventilation. Sequential ventilation can be used after the fire is out to clear contaminants out of any compartmentalized building. With the blower operating at the ventilation point and interior entry doors closed, firefighters should open up a door and exhaust point(s) in one section of the building. Courtesy Tempest Technology, Inc.

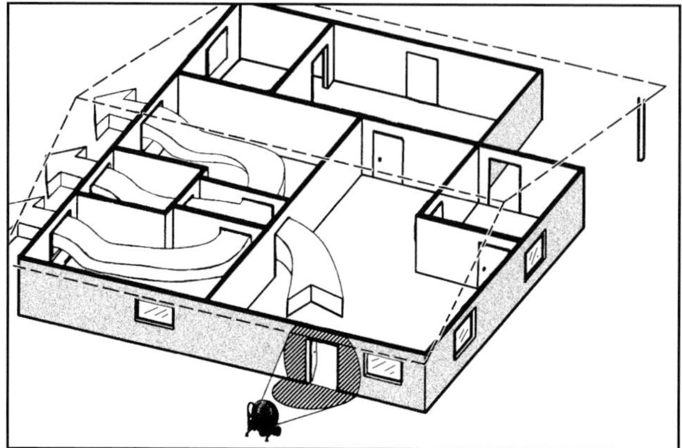

Fig. 7–10 Next, firefighters should close off the first area being ventilated. It is not necessary to close the exhaust point(s) as long as the access door is closed. Then they can open up another area the same way as the first. Courtesy Tempest Technology, Inc.

To clear hallways in a multistory building, with all other floors closed off, firefighters should clear the lowest floor first, then close it off and proceed to the next higher floor until all are cleared (fig. 7–11).

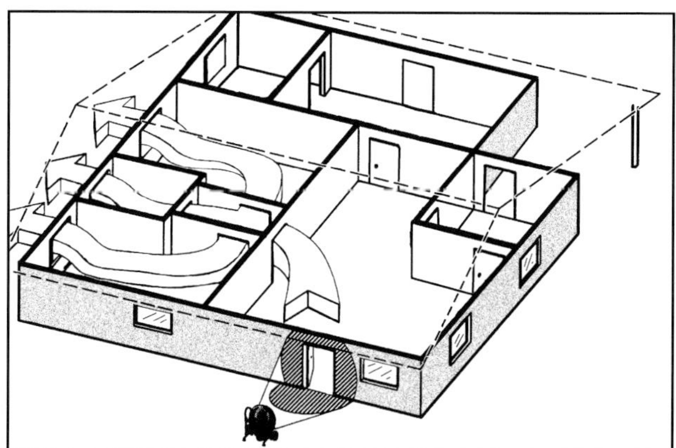

Fig. 7–11 Firefighters should repeat the ventilation operation as often as necessary in sequence to clear the building. Compartments may be individual rooms, groups of rooms, floors in a multistory building, or any combination. In a multistory building, firefighters should clear lower floors first. A pressurized stair shaft can serve as the ventilation point for each floor. Courtesy Tempest Technology, Inc.

Firefighters should remember that blowers will create elevated levels of CO in any interior operation. Care must be taken to protect interior crews or anyone else inside the building during this type of operation. SCBA should always be worn while gas-powered blowers are operating.

Large-scale ventilation

When faced with a large area to ventilate, something perhaps on the scale of millions of cubic feet, the latest innovation in mechanical ventilation is the mobile ventilation unit (MVU). Built by Tempest Technology, the MVU is designed to ventilate unique or very large areas such as warehouses, high-rises, aircraft, airport terminals, malls, and mile-long transportation tunnels. Models are available mounted on a motor vehicle chassis, on a trailer, skid mount, or rail mount.

The MVU can articulate 360° and elevates to provide flexibility and rapid pressurization. Although most structures can be adequately ventilated by the gas-powered blowers carried by many departments, the MVU offers the ultimate in control of the environment in massive areas (fig. 7–12).

Fig. 7–12 An articulating blower and elevating system makes the MVU a flexible option for pressurizing large areas such as warehouses, high-rises, tunnels, and, of course, aircraft. Courtesy Tempest Technology, Inc.

Pressurizing for Exposure Protection

Up to this point, positive pressure has been discussed as a method to stop the spread of a fire and remove heat and products of combustion. However, its versatility does not stop there. It has also proved to be a very valuable tool to help keep fire and smoke out of a building or area in the first place.

One of the authors recalls the following incident:

> *Mid-morning on a nice early fall day an alarm sounded for a countertop manufacturing warehouse and shop. Because of the heavy smoke showing, companies responding called for a second alarm while still en route blocks away. They arrived at a single-story building that measured approximately 100 ft by 300 ft and had a bowstring roof. On arrival the shop area in the rear was completely involved, and fire was threatening the office area in the front, as well as a large exposure to the north. The intense fire was fueled by sawdust and glue residue that had collected for years throughout the shop. Initial efforts focused on trying to keep the fire confined to the shop area.*
>
> *The second alarm companies were assigned to support the ladder pipe that arrived on first alarm. As the fire grew, crews started exposure protection operations. The south wall of one exposure paralleled the north wall of the fire building, a mere 2 ft away. The exposure was a single-story cinderblock building with an ordinary-constructed wood roof system that extended past the wall of the building approximately 1 ft. This put the wooden overhang within 1 ft of the fire building. The exposure had bulk paint storage with dozens of 50-gallon drums of solvent and other paint products that were stacked against the exposed wall.*
>
> *I was the captain of Engine 7 and was given responsibility for exposure protection. Firefighters from Truck 14 were already on the roof of the exposure, and other crews were directing master streams between the buildings. Because of the distance and intense heat from the fire, the master streams could not reach the last 50 ft of the exposure, and the roof crews could do little more than run water on top of the roof system.*
>
> *It became obvious that exterior coverage was not going to afford enough protection. I assigned my company to the inside of the exposure and placed a single blower at the front door to pressurize the interior.*

In the exposure, we kept windows and doors closed except for the ventilation point at the main entrance. Because it was such a large building, a second blower was placed inside to help pressurization.

Hoselines were taken inside the exposure to extinguish hot spots if any developed. Crews opened up ceiling areas and other voids along the wall adjacent to the fire building so they could observe and detect any fire extension into the exposure. From time to time, small areas would start to burn into the exposure. Crews monitoring the exposure in this smoke-free environment were able to prevent the exposure from becoming involved by applying small amounts of water on the fire in places where it tried to burn into the pressurized interior.

As heavy equipment was tearing down the fire building later that afternoon, the exposure building was open for business. Pressurization had helped keep out a very intense fire that burned within feet of it (fig. 7–13).

Fig. 7–13 Salt Lake City Engine 7, with one of the authors as officer, was assigned to help protect the exposed building on the right. They started by pressurizing it and then opened up ceiling areas and voids inside near the wall adjacent to the fire building to monitor fire extension. The exposed building was open for business later that same afternoon. Courtesy SLCFD

The one-two punch of exterior and interior hoselines and pressurization helped prevent the fire from burning into the exposure. This exposure was saved not only because of the hoselines outside but also because the elevated pressure inside discouraged fire from burning into the interior in places where it burned through from outside. It also afforded crews a clear environment in which to monitor the interior.

After search and rescue, one of the top priorities in a structure fire becomes the prevention of fire extension to other buildings and to adjoining areas in the same building.

Exposure types

Fire can be communicated to exposures in four ways: radiation, convection, conduction, and direct flame impingement. Exposures fall into two categories: exterior and interior.

Exterior exposures. This category includes structures or objects separate from the fire building that may be ignited. Exterior exposures could be adjacent to the fire or could be structures some distance away that could be ignited by burning embers.

Protection of exterior exposures requires a number of measures. This could include wetting down the building exterior. It also could include removing combustible plastic, wood, or fabric window curtains, coverings, and draperies, and closing heavy window coverings, as well as removing combustibles from exposed areas. Crews can be assigned to hoselines inside and outside to cool the building exterior, monitor conditions, and deal with fire extension as it occurs (fig. 7–14). All these are effective and still necessary when using pressurization. In the event fire does burn through an external barrier of an exposure, pressurization will discourage it from involving the interior and will maintain visibility for crews inside. Pressurization helps get the most from limited fireground resources.

Interior exposures. This includes separate areas within an involved building that are completely partitioned off from the burning area. Interior exposures can be horizontal, such as stores in a single-story strip mall, adjacent offices in an office building, and dwellings with attached garages (fig. 7–15). They can also be vertical, such as floors in a high-rise. In terms of pressurization for exposure protection, the outer partitions of interior exposures should be looked at as if they were exterior walls of a separate building.

Fig. 7–14 *Positive pressure and hoselines in the front and rear helped save this apartment building and the residents' belongings as a major fire raged just across the alley. Courtesy Martha Ellis*

Fig. 7–15 *Pressurization from the blowers at the front door of the dry cleaning shop in the foreground ultimately could not save it from being damaged by the fire running the open roof in the strip mall. However, it did keep smoke and fire from damaging the customers' clothing items, which were all safely removed. Courtesy Martha Ellis*

When outer partitions do not have openings that can become exhaust openings, pressurizing these areas will have no impact on fire operations in adjacent areas (fig. 7–16). It will also help keep the environment inside the exposure clear for firefighters to operate in, although CO could still build to levels where protection is required. Crews should place blowers at the street entrance or, when necessary, bring blowers inside the structure. These tasks are generally assigned to later-arriving companies, because the initial blower will be on the exterior of the building at the back of the initial attack crew to support interior fire attack. Pressurization can also help keep fire from extending upward in a multistory building.

Fig. 7–16 A V-cut in this overhead door provided quick access, and two blowers in series provided pressurization to protect the area from a major fire that destroyed an adjacent area of the same building. Courtesy Ray Schelble

An important note of caution here is that pressurization of interior exposures should always be closely coordinated with command. There should be no doubt that this is an exposure operation and not another point for attacking the fire. A breach made in a partition of an interior exposure could create a situation where two blowers are impacting the same space from different directions. This air movement could put attack crews in jeopardy, as could the effects of crews with opposing hoselines.

Pressurizing exposures

A blower elevates pressure in an enclosed area any time more air volume is forced in than the size of exhaust openings will allow to exit. But what happens if firefighters pressurize without any exhaust openings in a building that is not even on fire or in an uninvolved area of a burning building?

Pressurizing with no exhaust opening. In an area that has no exhaust opening, a blower raises the air pressure, but little or no air flows through except for an inconsequential amount that may leak through small gaps and holes in the structure. Used in this manner, the building interior reaches maximum pressurization. Although this use of pressurization will not efficiently remove contaminants from the interior atmosphere, it will help keep fire and smoke out.

In a pressurized exposure, fire can still ignite the outside of an exposed wall, so hoselines will probably still be necessary to protect the outside. But fire that does manage to burn through to the inside of the exposure instantly creates an exhaust opening. Fire burning right at the exhaust opening is a best-case scenario for positive pressure (fig. 7–17). The pressurization forces it to the outside, and the moving air cools it. Pressurization can help create a manageable barrier for crews faced with a fire that could extend beyond the building or area of origin.

Pressurizing exposures can be especially valuable in situations where resources, including water, equipment, and companies, are limited. A blower in the doorway of an exposed building can buy time for crews until later-arriving companies can take a more active role. Crews faced with protecting multiple structures with limited water supply or personnel should definitely consider pressurization as a means of helping to protect exposures.

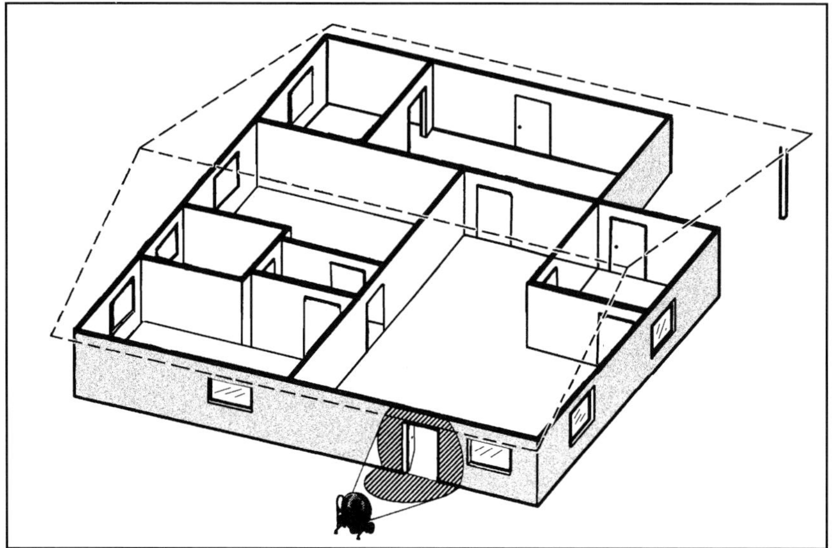

Fig. 7–17 To pressurize for exposure protection, firefighters should close up a building or compartment and direct the airstream from a blower run into the ventilation opening. Anywhere that fire does burn through from the outside becomes an exhaust opening, which keeps the fire outside. Courtesy Tempest Technology, Inc.

Procedures for exposure protection. In all situations where conditions allow, first-arriving crews should still use proper PPA in the area where active fire is burning. Except in certain circumstances, the first blowers at the scene should be used for PPA, and protecting exposures requires additional blowers.

Firefighters should set up blowers for exposure protection the same as for any positive pressure operation. The ventilation point on an exposure should be located away from the fire area so the blower will not push products of combustion, sparks, or flaming debris from the fire building into the uninvolved exposure. Ideally, the blower should be positioned 8 ft to 10 ft outside of the entrance to the area being pressurized, sealing the opening as much as possible with the cone of air.

In pressurizing for exposure protection, firefighters must be careful not to create exhaust openings. All exterior openings other than the ventilation opening should be kept closed to keep the interior pressure in the building as high as possible. Pressurization will still help protect an exposure that is not completely closed up as long as the openings are not too extensive.

Interior doors are another matter. Leaving interior doors open makes it easier to monitor conditions and ensures pressurization achieves maximum effectiveness in areas that may burn through. In critical areas that are at high risk of being ignited, such as ceilings or walls that are being impinged on, opening up void spaces for observation will help detect fire extension early enough to keep damage at a minimum.

Even though there will usually be very little air movement in this type of operation, small amounts of exhaust fumes and CO may still be introduced into the exposure. Either SCBA should be worn or CO concentrations should be monitored to ensure they remain within safe levels.

An additional tactic. Again, pressurization is not intended to replace hoselines and other means of exposure protection, and it will not prevent radiant heat through windows. It is an additional tactic to assist crews in keeping the fire from gaining ground inside an uninvolved space. Using blowers to prevent fire migration into a building takes little time to set up at the basic level. Even though pressurizing exposures works very well, interior conditions must still be monitored by firefighters with readily available hoselines.

Positive pressure can be a tremendous benefit to help protect exposures and can be used in most situations. However, it is not intended to take the place of the usual ways of protecting exposures or monitoring fire extension. Firefighters should consider it another tool to help them do their jobs better.

Pressurization by design

Fire protection engineers and architects have been using pressurization to protect vital areas of buildings and occupants from fire and smoke extension for more than two decades. When sizing up commercial or high-rise buildings, firefighters should consider that equipment to pressurize the interior may already be permanently installed as an integral system. In some cases old, negative pressure smoke removal systems in buildings have been reconfigured to provide exterior air for pressurization.

One of the most common uses of built-in pressurization can be seen in high-rise structures. In a building that is taller than ladders can reach, stairways become critical for the evacuation of occupants and for fire crews to mount a fire attack. The environment in a stairwell must be protected. Emergency pressurization systems built into heating, ventilation, and air-conditioning (HVAC) systems are common. Ventilation systems are often incorporated into elevator lobbies to protect these vital avenues of egress.

Pressurizing specific exposure types

Crews wanting to limit the spread of a fire should take steps to pressurize areas they determine are potential exposures. Although exposure protection follows the same basic principles, certain types of situations have unique characteristics that should be considered.

High-rises. As has been discussed, many high-rises and larger buildings already have ventilation systems that pressurize the building interior or stair shafts to prevent fire extension and the spread of products of combustion. Before starting pressurization in buildings that have smoke removal systems, firefighters should always determine the capability of the system. The building's engineer is a good source for information on how pressurization can best augment an existing ventilation system rather than risking ineffective pressurization that works against it. It is also wise to make note of these systems during preplanning. Any decision to use portable blowers for pressurization must take these systems into consideration.

When needed, the first blower on a high-rise fire should ideally be positioned 8 ft to 10 ft outside the exterior entrance to the stairwell that is going to be used by fire crews to access the fire floor (fig. 7–18). This keeps the stairwell clear while crews transport equipment upstairs and make their fire attack. At the same time, it makes it less likely that the fire will extend from the fire floor into that stairwell. An IC who is unsure of conditions on the fire floor is well advised to have crews investigate first. They can then advise on the best means of implementing stairwell pressurization, whether the purpose is to rid the stairwell of products of combustion or keep contaminants out of a clear stairwell. Additional blowers can then be set up at the entrance to the fire floor or in other locations as needed.

At the floor above the fire, blowers can be positioned in the stairwell outside the entrance door with the airstreams directed toward the interior of the floor. If the floor is charged with smoke, pressurization may force it to upper floors if there are openings between floors. If using another stairwell to exhaust a floor, firefighters must use caution that it does not interfere with evacuation or fire operations on upper floors. They must also ensure that all upper floors are closed off from the stairwell so they will not fill with smoke.

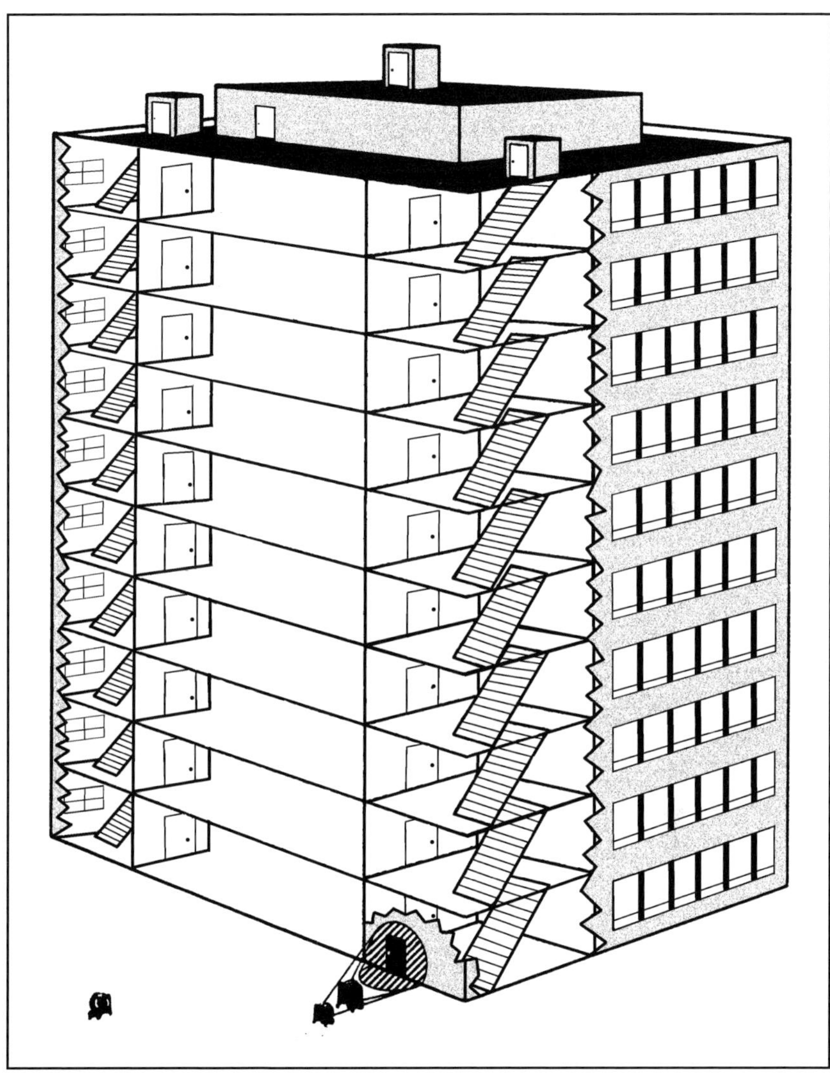

Fig. 7–18 Ideally, the first blower at a high-rise should be positioned 8 ft to 10 ft outside the exterior entrance that is going to be used by fire crews to access the fire floor. Courtesy Tempest Technology, Inc.

Fire attack and ventilation in a high-rise building are complicated issues. They must be addressed by aggressively deploying crews to the fire floor, supported by additional resources and tactics, including ventilation. All aspects, including doors and other openings, must be closely coordinated and controlled by a strong incident command structure (fig. 7–19).

More detailed information on ventilating and operating in high-rises will follow in a later section of this chapter.

Fig. 7–19 Fire attack and ventilation in high-rises are complicated issues that require a strong incident command structure. Courtesy Kriss Garcia

Strip malls. If the occupancy in question has adjacent exposures on both sides, a minimum of three blowers may be necessary. The first blower supports PPA for initial fire attack (fig. 7–20). The second and third blowers should be placed at access/ventilation points of exposures on both sides of the occupancy that is involved in fire. Keeping all other openings in these exposures closed maintains the pressure in the exposures above the pressure in the fire building, which discourages horizontal fire extension.

As crews become available, assignments should be made to each exposure to monitor the situation and stop any fire spread. As adjacent exposures are pressurized, the attic or cockloft area must be opened up to ensure fire is not spreading above the stores. If adjacent exposures contain high levels of products of combustion, openings may need to be made to the outside to clear the atmosphere to help crews working inside.

Fig. 7–20 For a fire in a strip mall occupancy, the first blower goes to support fire attack. Courtesy Kriss Garcia

Dwelling with attached garage. If there is any one type of building that takes advantage of all the positive impacts of PPA and pressurization, it has to be a fire in a garage that is attached to a dwelling. A fire in an attached garage can reap maximum benefits from all elements of pressurization with dramatic benefits to the attack crews and the homeowner. This is also a situation where fire attack and exposure protection can adequately be accomplished using one blower.

The first-arriving crew positions a blower at the entrance to the house for PPA and ensures an exhaust opening exists in the garage area. Attack crews then can quickly make their way through the house with hoselines and through the interior doorway to the garage, the whole time with pressurization at their backs. Pressurizing the dwelling keeps it free of smoke by treating it as an exposure (fig. 7–21). If the fire has not extensively invaded the living areas of the house, additional exhaust openings should not be made in the dwelling.

Fig. 7–21 A fire in a garage attached to a dwelling takes advantage of all the positive aspects of PPA and pressurization. Pressurizing the closed-up living area and attacking the fire through the connecting door between the house and garage provides both ventilation and confinement. Courtesy Kriss Garcia

This is a classic example of how a crew using PPA can effectively attack a fire and help protect an exposure simultaneously. Increasing the pressure inside the living space of the house keeps the fire in the garage. As the attack crew passes through the door from the house to the garage, the opening is pressurized as if there was a blower right at their backs. Operating a blower in this manner protects the house from smoke damage and fire extension, and assists with extinguishing the garage fire.

On the other hand, if crews decide they must quickly knock down the fire in the garage area with hoselines from outside, pressurizing the house first still greatly decreases the possibility that the fire will extend into the dwelling. It keeps the interior clear so that crews can operate in a safer environment while making a rapid advance on the fire.

In most instances, an attached garage fire can be quickly extinguished without unnecessarily compromising the life safety of fire crews or property. As an added benefit, the truck crew that may have otherwise been tied up with ventilation can now assist in forcible entry and search-and-rescue operations.

As with any significant structure fire, advancing crews should always check the ceiling area above them for fire extension.

Single-story, warehouse-type building. In a large, single-story building, whether it is a warehouse, mall, or mixed occupancy, solid interior partitions may be built to divide it into offices or storage areas. As with any interior exposure, these partitions can be considered as exterior walls.

For a fire that has self-vented on one end of a large building, firefighters should set up PPA at an entrance on an opposite end. This will not only start to ventilate the structure of heat and products of combustion, it will also limit fire extension.

As soon as possible, crews should take additional blowers into the structure and pressurize other interior partitioned areas not involved in fire that may be threatened by the fire, such as offices or stores (fig. 7–22). This will increase the pressurization in the fire area while also enhancing exposure protection in the building. All crews should wear full protective equipment, including SCBA, until CO concentrations drop to safe levels.

Fig. 7–22 Parallel blowers clear the interior after a fire in a temporary warehouse that was set up for the 2002 Winter Games in Salt Lake City. Courtesy Kriss Garcia

When fire control is ensured, additional blowers can be placed outside the structure to assist in complete ventilation of the building. Although CO levels will be elevated, they will not be as high as they would be if the fire extended into the building. The major benefit of limiting fire spread and operating in a clear and safe environment outweighs the minimal negative impact of introducing CO in the structure.

Exposures in wildland fires. Based on the authors' experience, the use of positive pressure to assist in protecting exposures during wildland fires has value under some circumstances. A building with impinging fire should be evaluated to determine whether pressurization could actually help. Pressurization will benefit an exposure only if the building's exterior has a chance of being protected. If an assault on the building from a large body of advancing fire or flying brands is likely, pressurization may not be of much help. Factors to evaluate when deciding whether to use pressurization include the extent of the impinging fire, exterior construction materials, status of water supply, winds, and availability of equipment and personnel (fig. 7–23).

Fig. 7–23 The decision whether to use pressurization to protect structures exposed to wildland fires depends on factors such as the extent of impinging fire, exterior construction materials, status of water supply, wind, and availability of resources. Courtesy Richard Moseley

Pressurizing to protect exposures in wildland fires is accomplished in much the same way as with other exposures. The blowers are placed 8 ft to 10 ft away from an opening on the side remote from the advancing fire, and the building is pressurized. The rest of the structure should be closed up as tight as possible.

If the advancing fire is likely to burn over or around the structure and ignite spot fires or create brands that the blower may push into the exposure, firefighters should not pressurize the building. If circumstances are such that the main risk of ignition is from a single advancing head of the fire, pressurizing may assist crews in protecting a structure, especially when equipment and personnel are scarce.

Ventilating High-Rise Buildings

The movement of products of combustion in a high-rise often creates a greater hazard to life and firefighting efforts than the fire itself. Smoke and hot gases may block escape routes and hamper fire suppression efforts. Heat and other products of combustion will most often remain in these buildings until forced out by carefully executed ventilation.

High-rise buildings present unique and varied ventilation problems to firefighters due to factors such as height (in the fire service, a high-rise is generally considered to be taller than 75 ft). Other factors include smoke movement characteristics, logistical challenges, and limited options to ventilate to the outside.

Lack of a specific plan for using personnel and equipment effectively could have grave consequences. Additionally, when ventilation operations are necessary, a clear understanding of high-rise building systems is essential.

High-rise size-up, ventilation, and fire attack are complex operations subject to many influences. This section is intended only as a very basic discussion about aspects that relate to ventilation and positive pressure.

Factors influencing smoke movement

The smoke movement characteristic called mushrooming occurs when heat and products of combustion are trapped inside a building. The hot gases rise until either they run up against a physical barrier or they stratify. When rising gases reach a physical barrier like a ceiling, they begin to spread out horizontally (fig. 7–24). In stratification, the hot gases rise in the structure, gradually cooling as they rise. Stratification occurs when they cool to the temperature of the surrounding air. They stop rising and stratify into layers.

Fig. 7–24 Hot gases rise inside a high-rise until they are stopped by a physical barrier or until they cool enough to stratify, then they start to bank downward. This is called mushrooming. Courtesy SLCFD

In both cases, as products of combustion accumulate, they tend to bank downward and spread horizontally. Mushrooming may occur several floors immediately above the fire floor or all the way to the upper floors. This depends on factors such as the size of the fire, the height and construction characteristics of the fire building, and the location of the fire floor. Mushrooming is especially common in open vertical spaces, such as stair shafts.

Most high-rises have some common avenues for fire extension and smoke movement built in. The list includes construction-related factors such as elevator and stair shafts, HVAC systems and unsealed openings, and several other factors not related to construction.

Elevator/stair shafts. Elevator and stair shafts may be located in a central core that runs the height of the building (known as a center-core building) or can be located in areas throughout the high-rise. Because they are vertical, these passageways provide excellent channels for smoke and heat to rise (fig. 7–25). To complicate matters, occupants and fire department personnel must use them for access and egress.

Fig. 7–25 Because they run vertically in a building, elevator shafts provide excellent channels for smoke and heat to rise. Courtesy Kriss Garcia

Elevators may service all floors of a building or they may service only selected floors, such as parking levels 1 to 4 (fig. 7–26). Banked elevators service a group of floors such as 1 to 15 or 15 to 30.

Fig. 7–26 Elevators may service all floors of a building or they may service only certain floors. Courtesy Kriss Garcia

Stair shafts may service all floors of a building, odd or even floors, or only selected floors. A stairway may or may not terminate on the roof, and may be equipped with a built-in ventilation system to keep it free of smoke in a fire.

HVAC systems. HVAC systems can form natural channels for distributing smoke throughout a building. These systems may operate manually or automatically and should always be used with the help of the building engineer (fig. 7–27).

They can also save the day when needed to ventilate a high-rise. Some systems can be adapted to handle emergency ventilation, and others have emergency ventilation systems built right in.

Unsealed openings. Most high-rise buildings "leak." Products of combustion may move from floor to floor and into stair shafts through unsealed openings made for air ducts and electrical components, pipe chases, and other openings between floors (fig. 7–28). A fire can generate pressure up to three times atmospheric pressure inside a sealed building, which forces products of combustion through these leaks.

Other factors. In ventilation and the spread of products of combustion, much depends on the building's construction features and whether crews can use these features to advantage. But other factors come into play as well, such as doors, wind, and open windows and auto exposure.

Fig. 7–27 Building HVAC systems may be operated manually or electronically. In an emergency, they should always be operated with the help of the building engineer. Courtesy Kriss Garcia

Fig. 7–28 If left unsealed, spaces around air ducts, electrical components, pipe chases, and other openings can create "leaks" between floors. Courtesy Kriss Garcia

Doors. Products of combustion drawn into stair shafts through open doors may create an exposure problem that affects anyone using the stair shaft and could contribute to heat and contaminants spreading to upper floors. On the other hand, closed doors can isolate an area that should be ventilated (fig. 7–29).

Fig. 7–29 Firefighters must carefully plan which doors to keep open and which to close, as this will affect whether ventilation is effective or creates an exposure problem. Courtesy Kriss Garcia

Wind. Wind generally will not affect air movement much in an intact, sealed building. Most high-rise buildings with external glass panels that cannot be opened are classified as "sealed" buildings. In a sealed building, the internal environment, including such things as temperature and humidity, are all controlled with HVAC systems. These buildings retain smoke, heat, and fire gases until they are deliberately ventilated to the outside or until windows fail due to exposure to fire.

But wind can have a drastic effect on smoke movement and ventilation, especially when a floor needs to be horizontally vented outside. Conditions at the local weather forecasting center or even on the ground at the fire building may be drastically different from conditions several floors above street level (fig. 7–30). Opening up the floor below the fire floor is the most dependable method of determining wind direction on the fire floor. As with other ventilation situations, firefighters should use wind to advantage.

Fig. 7–30 Wind conditions on the upper floors of a high-rise may be drastically different than on the lower floors. Courtesy Ray Schelble

Open windows and auto-exposure. Exterior windows that have been opened or broken create uncontrolled horizontal currents that can direct smoke in unpredictable directions. In some cases, this can serve to help ventilate the interior. In other cases, fire burning out of these windows and up the outside of the building can impinge on upper-floor windows. If these upper windows fail, it creates a new path for fire extension and for products of combustion to spread (fig. 7–31). This process is called auto-exposure. Auto-exposure also depends on factors, such as fire load, flame height, and building design, that affect a building's vulnerability to this type of fire extension.

Fig. 7–31 Fire burning out of windows and up the outside of a building can spread fire to upper floors in multistory buildings. Courtesy Kriss Garcia

Tactical considerations

Due to the complexity and hazard potential of high-rise buildings, the importance of prefire planning cannot be overemphasized. As much information as possible should be known before the alarm sounds (figs. 7–32 and 7–33).

Fig. 7–32 *The importance of prefire planning cannot be overemphasized at high-rise incidents. These firefighters wait for the okay to start ventilation as crews investigate a fire on the 23rd floor. Courtesy Kriss Garcia*

Fig. 7–33 *The blower from the previous photo was put in service on the first floor to pressurize the stairwell. Smoke from the 23rd floor was vented into this stairwell and exhausted through the penthouse door at the roof. Courtesy Kriss Garcia*

Firefighters should take three important steps when sizing up a fire in a high-rise. These are: (1) verify the location and extent of the fire; (2) identify stair shafts and elevators; and (3) initiate systematic ventilation.

Verify location and extent of the fire. Because of the way they are built, most high-rises are capable of keeping extensive fire and smoke hidden with little or no visible signs outside. The first step is to verify there is a fire. If there is a fire, firefighters must determine the exact location and extent of the problem, as well as the area affected by smoke and products of combustion. Crews involved in the initial size-up must exercise extreme caution and be prepared for anything when investigating and establishing safe access.

If the building has a fire control system, firefighters should go to the control center (fig. 7–34). They should check the heat and smoke indicators for the location and extent of the problem and identify the designated fire department elevators. They also should contact the building engineer or maintenance personnel to mobilize the building's HVAC and integrated smoke removal systems.

Fig. 7–34 A building's control center provides a valuable resource to crews investigating a fire in a high-rise. Courtesy Kriss Garcia

Determine locations of stair shafts and elevators. A building's stair shafts and elevators are of major strategic importance. Their status and termination points must be determined, preferably during preplanning.

Stair shafts. A properly managed stair shaft can provide a natural channel to ventilate smoke and fire gases in a high-rise. Some stair shafts have built-in pressurization fans that can be activated to create an upward flow of air when a door is opened at the top. These integrated fans are rated well below fire service blowers in terms of volume, sometimes moving only one-half the volume.

If necessary, PPV can be combined with these built-in pressurization systems to increase airflow and efficiency. Some stair shafts also have automatic or remote-operated dampers that control venting to the outside and can be operated from a safe location such as a fire control room (fig. 7–35). Stair shafts that open to the roof can quickly and easily be cleared of contaminants, affording both firefighters and occupants safe access and egress.

Fig. 7–35 Stair shafts may have automatic or remote-operated ventilation systems, such as this one on the roof of a stairway structure, that can be used to advantage when ventilating. Courtesy Kriss Garcia

In stair shafts that do not have a pressurization system, positive pressure using fire department blowers can effectively discharge products of combustion. To pressurize a stairway, blowers should be positioned at the ground-level entrance door outside. If the stair shaft does not open to the outside, blowers should be positioned at the building entrance and at the interior entrance to the stairway.

Firefighters must determine the best stair shaft for the initial crew to use to investigate the extent of the fire. To avoid confusion, when possible, they should separate and identify stair shafts used for ventilation and fighting the fire from those used for egress by building occupants.

Elevators. Firefighters must determine if the building has a service elevator, an elevator designated for fire department use, or banked elevators. Service elevators usually travel the full height of the building and can provide access to all floors, including loading and freight docks (fig. 7–36). Fire department elevators offer manual control, which can provide additional flexibility. Firefighters should check whether the passenger elevators travel the full height of the building or if they are banked. Elevators should not be used until their safety has been verified by the investigating crew. The first-in crew should use only banked elevators for lower floors or stairways for access. The safety status of an elevator may change at any time during the incident.

Initiate systematic ventilation. When investigating crews have verified there is a fire or a situation calling for ventilation, a decision needs to be made on the best way to ventilate. There are several options to consider for removing smoke and preventing its spread to unaffected areas. These are vertical ventilation through stair shafts, cross-ventilation, HVAC systems, and compartmentalization.

Fig. 7–36 Service elevators usually travel the full height of a building, while banked elevators service only certain floors. Courtesy Kriss Garcia

Vertical ventilation through stair shafts. Firefighters may use fire department blowers, built-in stairway pressurization systems, or both to remove contaminants and keep them from reentering a stair shaft once it clears (fig. 7–37). Opening the roof access with PPV in operation allows the contaminants that do enter the stairwell to be forced upward and out.

Fig. 7–37 Built-in ventilation systems, fire department blowers, or a combination of both can be used in a stair shaft to exhaust contaminants at the roof. Courtesy Kriss Garcia

As a general rule, in stair shafts taller than 25 stories, additional pressurization will be required in order to overcome the loss of air through vents and other unsealed openings. When the stair shaft is used to remove contaminants from floors above the 25th story, additional blowers may be necessary to assist in pressurization.

Doors play an important part in any attempt to ventilate a high-rise and should be left open only as part of a systematic plan. Firefighters must pay special attention to stairwell doors.

Simply opening a door at the top and bottom of a stair shaft can quickly create a natural upward flow of air. This can be exploited to remove smoke and fire gases from the stair shaft or from the fire floor. For ventilation purposes, stair shafts fall into two categories: interior access and exterior access.

Entry to interior access stair shafts is from the inside of the building at the bottom and from the interior floors. These stair shafts open outside onto the roof. Mechanically-assisted ventilation, either using the building's ventilation system, fire department blowers, or both, is necessary for emergency ventilation.

Exterior access stair shafts have access outside of the building at the bottom and to the interior floors. At the top they open outside to the roof. Unlike an interior access stair shaft, the natural vertical airflow can reach speeds of 3 mph to 6 mph simply by opening doors at the bottom and at the roof. This can be enough air movement to remove contaminants from a fire. As with interior access stair shafts, the natural vertical airflow in these can be assisted by using integrated smoke removal systems alone or in conjunction with PPV (fig. 7–38).

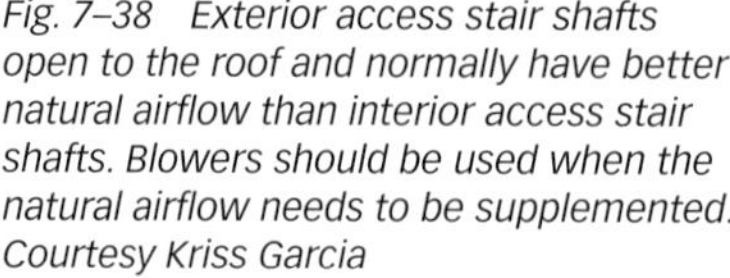

Fig. 7–38 Exterior access stair shafts open to the roof and normally have better natural airflow than interior access stair shafts. Blowers should be used when the natural airflow needs to be supplemented. Courtesy Kriss Garcia

Cross-ventilation. A smoke-filled floor in a high-rise building can be ventilated horizontally (cross-ventilated) through windows with blowers or a combination of blowers and stair shaft pressurization fans. If possible, smoke and gases should be vented through windows on the downwind (leeward) side.

When windows are not available or cannot be opened, a stair shaft on the opposite side of the building that opens to the roof will also work. Pressure from the upwind stair shaft can be directed through the fire floor and exhausted vertically through the downwind stair shaft and out the roof door.

To ventilate sealed buildings, crews may need to remove or break panels or windows in the exterior walls. It is important to always communicate with the crews outside of a high-rise before breaking any windows in the upper levels.

HVAC systems. Many HVAC systems include provisions for fire ventilation. Some systems have remote-controlled dampers at the supply and return ducts that close to prevent heat loss when the building is unoccupied and open to supply conditioned air to the building. In an integrated and designed HVAC system, these can also be used to control smoke.

Many variations are found in emergency smoke control systems. Some systems are manually or automatically activated by smoke detectors and alarms, while others use a combination of manual and automatic features.

When facing a threat of deteriorating fire conditions, firefighters should shut the HVAC system down and observe the effects on the fire situation. Other than shutting a system down completely, HVAC systems should be controlled only by the building engineer or someone familiar with them. It is better to err on the side of caution than contribute inadvertently to the spread of fire or contaminants.

Compartmentalization. Pressurizing floors that have not been contaminated by smoke and keeping them closed off from other floors will discourage fire and products of combustion from spreading into them through ventilation systems or construction leaks (fig. 7–39). Simply setting up a blower inside a stairwell that opens to the outside and directing the airstream into the entrance to the floor may be enough. As in other applications, the ideal positioning for the blower is 8 ft to 10 ft from the doorway, but this is flexible. Multiple blowers may be needed to pressurize larger areas, especially when a building has abundant air leaks. Another blower at street level to pressurize the stairwell will increase the efficiency of the operation.

Fig. 7–39 Individual rooms create compartments in this hotel. Compartmentalization can be used to advantage in high-rises as in other types of buildings. Uninvolved areas should be closed off to keep fire and contamination out and ventilate sequentially. Courtesy Kriss Garcia

No exhaust opening should be made unless there are contaminants to be removed. Compartmentalizing in this way can reduce fire loss by minimizing damage from smoke and heated gases on floors above the fire. For safety's sake, firefighters should never go above a fire floor without a safety line or hoseline. Regardless of the operation, all companies should be working within a systematic plan, implemented by a trained, knowledgeable IC.

Firefighters must start somewhere when learning any new firefighting method. In the next chapter, the authors share some of their experiences and observations on how to introduce positive pressure to firefighters and department administrators.

8

Introducing PPA to the Department

I hate quotations. Tell me what you know.

—Ralph Waldo Emerson

The background information has been presented, and now it is time to get to the details of implementing PPA department-wide. The fire service covers a range of firefighters from career to volunteer, with various combinations in between. These men and women are absolutely committed to providing the best service possible to their communities, whether in a rural area or a major metropolis.

Firefighters everywhere are inquisitive. This is obvious to anyone who reads any of the popular fire service journals. Firefighters like to know what other departments are doing, how they are training, what types of apparatus they are using, where they stand on their labor-management issues, and who is in or out of the chief's office. And it is certain that firefighters will look carefully before accepting positive pressure.

The SLCFD Experience

In early 1991, after some lobbying by the authors, the Salt Lake City Fire Department started with one blower on one engine. The acceptance of it was certainly in question. On two platoons, the officers would keep the blower on the apparatus. On the other platoon, the officer would remove it because he felt the compartment space was better utilized to store his bunker gear. It was obvious that a great amount of work had to be done before firefighters would accept this new form of ventilation. It also became clear that firefighters needed proof rather than someone telling them of other departments' success or showing them data and statistics that proved that pressurized attacks were more effective. They wanted firsthand knowledge and experience that demonstrated the principles of PPV and, later, that PPA actually worked.

Fluttering toilet paper

The authors began proving their case with the most rudimentary equipment and knowledge—blowers, toilet paper, and the theory that air will move from areas of high pressure to areas of lower pressure. Taping strips of toilet paper to doors and windows showed where air was moving and in what direction it was being directed when blowers were turned into the building. From this, firefighters could visualize smoke moving out of a structure by observing the toilet paper flutter (fig. 8–1). They could practice directing air replacement in the interior environment by opening and closing doors and selecting different exhaust openings. It was an effective first step, but not enough to demonstrate all of the advantages of PPV to skeptical firefighters.

Fig. 8–1 In the early 1990s, the authors used fluttering toilet paper and other tools to research positive pressure. Notice the toilet paper blowing horizontally at the ventilation point in this still shot from an early video as Reinhard checks the air velocity. Courtesy SLCFD

Since PPV is all about removing smoke, what could be a more natural second step than getting a smoke machine to fill up a structure (sometimes a fire station)? Firefighters would then utilize blowers to clear it out. They were able to watch PPV quickly force smoke out of the structure and see firsthand the advantage of increased visibility the blowers provided.

The next step

"All right, we can make toilet paper flutter and exhaust cold, simulated smoke out a building. So what?" These firefighters were still not convinced that PPV really provided any significant benefit over the traditional ways they had been ventilating structures. In fact, they questioned if the use of blowers was safe at all in real fire situations, since "everyone knows that blowers increase the size of the fire and spread it through a building." They certainly had a valid point that it needed to be proven under fire conditions.

But how could the theory be safely tested under live fire conditions? The first live fire tests were conducted in the burn room at the SLCFD training tower, which could handle real fire in a safe, controllable environment. Fire after fire in the burn room produced some stunning conclusions:

- The air movement produced by the blowers did not create velocity inside the burn room, only at the ventilation opening and at the exhaust opening. This demonstrated that blowers would not push or spread the fire.

- The use of blowers during the free-burning stage actually reduced the temperatures inside the room, a good thing if one wants to prevent flashovers and increase victim survival. Blowers also reduced the intensity of free-burning fires in these tests.

Going inside

After some initial tests, firefighters entered the burn room during live fire tests and were able to observe firsthand as heat and products of combustion were quickly and effectively ventilated by the blowers. They also witnessed other benefits, including increased visibility, reduced temperatures, and the ability to put water on the fire quickly (fig. 8–2). Timely ventilation made it easy to get to the seat of the fire.

Fig. 8–2 One of the first steps to test PPA was in SLCFD's Class A burn room. Here, participating firefighters witnessed flame and heat impressively exhausting from the exhaust point, while the inside of the room quickly cleared of heat and smoke. Courtesy SLCFD

One major observation made during these burn room fires was that the large volume of heat and products of combustion being forced out through the exhaust opening did not reflect at all what was happening inside. Firefighters were able to actually see the two different environments. On the exterior they saw flames and smoke being exhausted under pressure, while on the interior they saw a significantly improved and tenable environment. Many of the skeptics became more interested after witnessing these experiments firsthand. However, others still refused to believe in their hearts what their eyes had witnessed.

"Test driving" in buildings

After these tests, the authors continued to feel strongly that utilizing positive pressure as an initial attack option held great promise. They moved to the next stage: live fires in acquired structures. If one really wants to see what happens with any new product or idea in the fire service, there is no better "test drive" than repeated fires in abandoned structures.

In 1994, the first series of acquired structure burns took place in a single-story, wood frame dwelling. These burns were designed to accomplish several goals: to observe and document PPV, to determine whether fire is spread or pushed through the structure, and to experiment with using PPV on attic fires. Firefighters who rotated through the exercises noticed and commented on the most striking differences in the interior attack. They were able to make a quick entry, and they could walk through the building standing up as smoke and heat cleared before them. In addition, they could clearly see the seat of the fire, and they were not punished by high temperatures. During the course of these burns, fires were started in every room, and all the outcomes were positive and consistent with what has been presented previously in this book (figs. 8–3, 8–4, and 8–5).

Fig. 8–3 Figures 8–3, 8–4, and 8–5 are frames captured from various stages of a 1994 video made by the authors documenting a PPA test burn in an acquired building. Two fires were started, one in a room adjacent to the ventilation point and another in the attic. In this frame, before starting the attack, the fire can be seen through the ground-floor window on the right as it rages inside. The front door entrance is located at the left center of the picture. Courtesy SLCFD

Fig. 8–4 In the second frame from the 1994 video, the blower is turned on, and intense fire belches from the ground-floor window and from the opening at the peak of the attic as they both become exhaust openings. Courtesy SLCFD

Fig. 8–5 In the third frame, the heat and products of combustion forcefully vent from ground-floor windows near the entry and on the right, but firefighters inside find an atmosphere that is surprisingly cool and clear. Even though the main-floor fire was started right next to the ventilation point and the attic is obviously opened up, pressurization did not move fire through the building, and the attic fire actually diminished in size. Courtesy SLCFD

In 1996 a second series of live burns had firefighters arrive on their apparatus, secure a water supply, set up blowers, stretch hoselines, and a make an interior attack (fig. 8–6). These test burns were conducted in several single-family structures that were identical in construction. This afforded an opportunity to replicate burns under similar conditions and evaluate different blower scenarios. These burns were the first experiments with PPA.

Fig. 8–6 In a 1996 live burn, a crew on the roof works to create a vertical vent hole while another crew attacks using PPA. The flames exhausting from the window indicate that ventilation is already underway while the roof crew still works. Courtesy SLCFD

Acceptance evolves

At the conclusion of these burns, the authors noticed a definite shift in the firefighters' acceptance of using the blowers in fire attack evolutions. After repeated exposure to the benefits of this new way of achieving a coordinated attack, the firefighters were beginning to realize the benefits that positive pressure could provide.

The department administration was convinced at this point. PPA evolutions were developed and included in the department's firefighting evolutions manual, and blowers became standard equipment on all apparatus (fig. 8–7). The more the department's crews used the blowers, the more comfortable they became with the idea. Ventilation was finally becoming coordinated with initial fire attack.

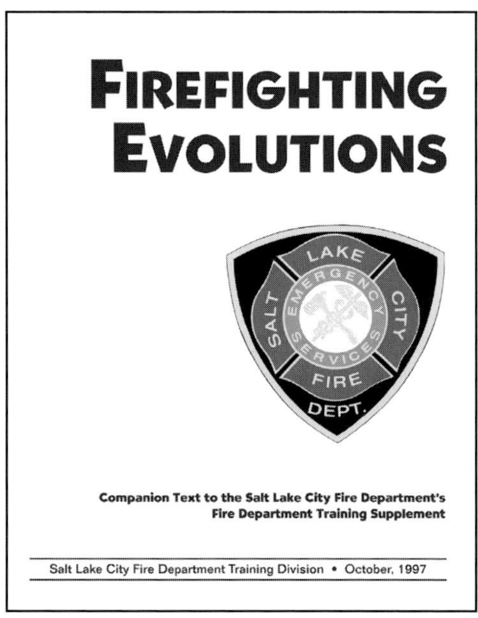

FIREFIGHTING EVOLUTIONS

Companion Text to the Salt Lake City Fire Department's Fire Department Training Supplement

Salt Lake City Fire Department Training Division • October, 1997

Fig 8–7 Once department supervisors were convinced, PPA evolutions were developed and included in the SLCFD's firefighting evolutions manual. Blowers became standard equipment on all apparatus. Courtesy SLCFD

As time passes, more and more officers and crews are feeling comfortable with PPA. As those who learn the method teach their crews, more and more firefighters are becoming convinced. It took several years, but momentum is now increasing. There is no stopping such an effective and proven method. Crews continue to fine-tune the operation to obtain the full benefits for their firefighters and those they serve.

Salt Lake City's Teaching Methods

The SLCFD's instruction and training in PPV/PPA begins with recruit firefighters during the department's training academy. Students need a healthy dose of facts and studies that prove PPA works, but more importantly, they need an experience that grabs them both intellectually and emotionally.

The intellectual side

On the intellectual side, they see a demonstration of the time it takes to set up a roof operation versus PPA. The authors show a video of a truck company arriving at a scene and performing a real-time, tactically correct roof ventilation evolution. While the video plays, simulated smoke is pumped into the classroom. As the smoke thickens, the students must come closer and closer to the television to be able to continue to watch the truck operation.

When the smoke eventually becomes too dense even to see the video, the students are asked if they think it is about time to begin ventilation, and if victims could survive this environment much longer. A window is then opened and a blower, which is positioned outside the building, is started. This rapidly clears the smoke, just as predicted. After less than 1 minute, the room clears, but the truck crew on the video is still working on the roof hole. While the truck company on the video is still working to open the roof, the classroom has been completely cleared of smoke, allowing search and rescue and effective extinguishment to begin and end before the ceiling has been poked through for roof ventilation. It takes the roof crew on the video almost 15 minutes to finish the task. Following this demonstration, the lecture portion of the class discusses much of the subject matter contained in previous chapters (fig. 8–8).

Fig. 8–8 After a thorough classroom orientation, SLCFD recruits receive hands-on training at the department's state-of-the-art training facility. Courtesy Ray Schelble

The emotional side

Just when the students are feeling comfortable with the concepts, their training continues by tapping into the emotional side of their learning experience. The experience with the fatal house fire that resulted in the death of the child is retold to them to drive home the point that PPA evolved from the desire to increase chances of victim survival. It is explained that to accomplish that goal, a change must occur in how fire departments operate.

Recruits then participate in live fire ventilation scenarios at the state-of-the-art SLCFD training facility. This includes scenarios for ventilating attached garage fires, room and contents fires, and high-rise fires. They receive hands-on training in proper blower placement, ventilation and exhaust openings, and clearing structures of smoke and heat by systematically opening and closing doors and windows. The goal is to give them enough information so that when they go out to the stations, they are well trained on the fundamentals and theory of PPA. The seeds have been planted.

Young firefighters are excellent prospects to teach new methods, especially when they can be shown something obvious that makes sense. They are easy to instruct because they are a captive audience.

Firefighters on the street can be given the same information, but it must be in shortened classes. The SLCFD's success in having firefighters use PPA is directly related to their having witnessed it work at actual fires. After a while they often offer comments like, "You know, those blowers really do work!"

Blowers are now routinely utilized by the department's fire crews (fig. 8–9). The overwhelming reason why is not because of the great instruction they get, but because positive pressure really works. It speaks for itself if given the opportunity.

Fig. 8–9 Positive pressure blowers are now routinely used not only by Salt Lake City firefighters but also by other departments in the area. Courtesy Richard Moseley

The next chapter explains in detail what each crew member has to do to put PPA in operation at an incident, based on firefighting evolutions that the SLCFD has developed.

9

Positive Pressure Evolutions

It has already been said, but it is so important it needs to be said again: Firefighters who think they can simply put a blower on their apparatus and begin to provide coordinated, safe PPA are sorely mistaken. Untrained firefighters who go out and start using PPA pose a danger to themselves, those around them, and the citizens they protect. Complete understanding of the nuances of operating in a pressurized environment is essential. Effective PPA requires knowledge, training, teamwork, efficiency, and a strong IC.

Ensuring Competency

It is vital to ensure that ventilation and fire attack are conducted in a coordinated and systematic way. To do so, a fire department must have an operational plan that its firefighters can follow and refer to when practicing and deploying PPA at fire emergencies. These systematic fireground routines, or operating procedures, are what the authors term *evolutions*. Evolutions must be developed with consideration for each department's capabilities and needs. Most importantly, they must be adopted and accepted by the department (figs. 9–1 and 9–2).

Fig. 9–1 Figures 9–1 and 9–2 are SLCFD firefighting evolutions that provide standard guidelines for implementing PPA. Evolution 1 details PPA for a four-firefighter crew on a first-in engine that is able to attack a fire without connecting to a hydrant. Courtesy SLCFD

Adopted 10/97

Ventilation Evolution #1

Positive Pressure Attack (PPA) with fast attack

Purpose: To increase the effectiveness of a fast attack on a fire by implementing PPA.

Description: The apparatus arrives at the fire scene and goes in with a fast attack using a 1-3/4 inch line supplied by tank water and PPA. Evolution begins when apparatus stops at the scene and ends when PPA is in place and the attack line is flowing an effective stream and advanced inside.

Notes: ❑ Safety and common sense take priority in all fireground activities. Use all required safety equipment, including full PPE where needed. ❑ Do not use PPA if a backdraft condition exists. ❑ Allow blowers to operate for 30 seconds before advancing the attack line inside. ❑ The entrance opening must be kept clear—firefighters standing in the entrance opening can seriously reduce the effectiveness of PPA. ❑ Apparatus operated with headlights on. ❑ Open and close gate valves slowly and carefully with two hands. ❑ Seat belts must be used when seated on apparatus. ❑ Straddling hose or hose clamps prohibited at all times.

Position	Procedures *(Also see the chapters, "Fire Attack" and "Ventilation," in the SLCFD Training Supplement.)*
Officer	❑ **Assigns tasks:** Notifies crew and incoming companies of the decision to go in on a fast attack and that a later-arriving company may need to lay a supply line. After determining a backdraft condition does not exist, gives the order to initiate PPA with fast attack. Advises hydrant/fan firefighter on the placement of the fan. Informs nozzle firefighter to pull a 1-3/4 inch pre-connect attack line. ❑ **Sizes-up building:** Grabs an appropriate tool and sizes-up the building. If necessary, makes or improves on a horizontal vent at a location near the fire. ❑ **Supervises and assists:** Returns to crew and supervises PPA and the interior operation. Helps gain entry through the entrance opening and assists with the fire attack.
Nozzle	❑ **Pulls attack line:** As directed by the officer, pulls the appropriate attack line and advances it to the building entrance used for the fire attack. ❑ **Entry and attack:** If necessary, helps gain entry through the entrance opening. Begins fire attack after the blower runs for 30 seconds.
Hydrant/ Fan	❑ **Positions blower:** Secures the blower and a forcible entry tool from the apparatus. Extends the handle on the fan and wheels it to the point of entry as directed by the officer. ❑ **Operates blower:** When attack line is in position and ready, starts the blower and positions it properly in the entrance opening. Ensures the opening remains clear.
Driver/ Operator	❑ **Sets-up apparatus:** Positions apparatus. Before leaving cab, sets parking brake, engages pump, and puts apparatus into proper gear. ❑ **Safety:** Places chock under apparatus wheel, and positions safety cones to warn traffic. ❑ **Charges attack line:** Opens tank drop. Slowly charges the attack line by partially opening the discharge valve until the line is filled, then opens the valve to provide an effective stream.

Comments:

Fig. 9–2 SLCFD's Evolution 2 lists PPA steps for a four-firefighter crew on a first-in engine that connects to a hydrant. Training evolutions should be developed with consideration for each department's capabilities and needs. A copy of these evolutions is given in appendix B of this book.

Adopted 10/97

Ventilation Evolution #2

Positive Pressure Attack (PPA), taking a hydrant

Purpose: To increase the effectiveness of a fire attack requiring water supply from a hydrant by implementing PPA.

Description: At the hydrant — The crew takes a forward lay from the hydrant with LDH or 2-1/2 inch hose following procedures in Hose Evolution #1 or Hose Evolution #2, then drives to the fire scene. At the fire scene — The crew completes the hydrant connections and attacks the fire using PPA and a 1-3/4 inch or 2-1/2 inch attack line. Evolution begins when apparatus stops at the fire scene and ends when the attack line is flowing an effective stream and advanced inside.

Notes: ❑ Safety and common sense take priority in all fireground activities. Use all required safety equipment, including full PPE where needed. ❑ Do not use PPA if a backdraft condition exists. ❑ Allow blowers to operate for 30 seconds before advancing the attack line inside. ❑ The entrance opening must always be kept clear—firefighters standing in the entrance opening can seriously reduce the effectiveness of PPA. ❑ Apparatus operated with headlights on. ❑ Open and close gate valves slowly and carefully with two hands. ❑ Seat belts must be used when seated on apparatus. ❑ Straddling hose or hose clamps prohibited at all times.

Position	Procedures *(Also see the chapters, "Fire Attack" and "Ventilation," in the SLCFD Training Supplement.)*
Officer	❑ **Assigns tasks:** *At the hydrant* — Directs crew to take a hydrant with LDH or 2-1/2 inch hose following procedures in Hose Evolution #1 or Hose Evolution #2. *At the fire scene* — After determining a backdraft condition does not exist, informs crew that PPA will be initiated and directs nozzle firefighter to pull a 1-3/4 inch or 2-1/2 inch attack line. ❑ **Positions blower:** Secures the blower and an appropriate tool from the apparatus. Extends the handle of the fan and wheels it near the point of entry ❑ **Sizes-up building:** Carrying an appropriate tool, sizes-up the building. If necessary, makes or improves on a horizontal vent at a location near the fire. ❑ **Assists with attack:** Helps gain entry through the entrance opening and starts the blower if it is not already running. When attack line is in position, assists with the fire attack.
Nozzle	❑ **Pulls attack line:** As directed by the officer, pulls the appropriate attack line and advances it to the building entrance used for the fire attack. If necessary, attaches a hose strap. ❑ **Entry and attack:** Helps gain entry through the entrance opening. Assists with blower. Begins fire attack after the blower runs for 30 seconds.
Hydrant/ Fan	❑ **Assists with attack:** After taking the hydrant following procedures in Hose Evolution #1 or Hose Evolution #2, goes to the objective with forcible entry or overhaul tools in hand.
Driver/ Operator	❑ **Sets-up apparatus:** Positions apparatus. Before leaving cab, sets parking brake, engages pump, and puts apparatus into proper gear. ❑ **Safety:** Places chock under apparatus wheel, and positions safety cones to warn traffic. If necessary, places hose clamp on the supply line, on the hydrant side of the last coupling. ❑ **Gets water into pump:** Begins operation using tank water if appropriate. Sees that the supply line is connected to the pump intake. Removes hose clamp if necessary. After the supply line is charged, slowly opens the gate into the pump intake. Changes-over from tank water to hydrant water if necessary. ❑ **Charges attack line:** Slowly charges the attack line by partially opening the discharge valve until the line is filled, then opens the valve to provide an effective stream.

Comments:

Evolutions assign predetermined tasks to each crew member on an apparatus, which ensures that all necessary assignments are completed and that crew members all work together as a team. All firefighters and ICs must train on these tasks until they have demonstrated competency. Only then, and with an understanding of positive pressure fundamentals and safety issues, can PPA be deployed in a safe, systematic, coordinated manner without adding time to the attack. The only way to take advantage of PPA to make timely, survivable rescues and fire stops is to practice, practice, and practice (fig. 9–3).

Fig. 9–3 Crews with the Langley Fire Brigade, British Columbia, work on their PPA skills and teamwork on the training ground. Training is the key to learning PPA. Courtesy Reinhard Kauffmann

The authors' experience has shown that crews that are well trained using PPA evolutions are much more likely to utilize blowers in their initial fire attack. The more competence crews gain on the training ground, the more confidence they will have using it on the fireground. Crews that have not gained proficiency are much more likely to leave the blower on the apparatus and not realize the maximum benefit it can provide. These crews will deploy attack lines and call for later-arriving units to provide ventilation.

Staffing, equipment, and other resources differ from one department to another, so the best way to systematically put PPA into operation may differ as well. Consequently, the most practical evolutions for each department may vary from the ones discussed here.

In looking at these evolutions, it is helpful to view them in the light of some important information that was presented earlier—the definitions of *coordinated* and *systematic.*

- *Coordinated* means "to act together in a smooth, concerted way."[1] With PPA, firefighters on the first-arriving apparatus are capable of coordinating hoselines and ventilation in one routine, or evolution, that does not add time to the initial attack.

- *Systematic* means a practiced and proven method of deploying this tactic for fire attack. Firefighters who are well trained will begin systematic PPA at a majority of their fire emergencies.

PPA Evolutions

These two PPA evolutions, developed by the Salt Lake City Fire Department and now utilized in many other departments, are included here and in appendix B of this book as examples of how PPA can be effectively put into practice. The only difference between the two is the decision of the first-arriving company to secure a water supply. The first evolution assigns tasks for fast attack on the fire. The decision to go in without connecting to a hydrant should be made only when the first-in engine has adequate water onboard or when other responding units are sure to arrive in time to provide needed water. The second evolution is for fire attack when a hydrant must be taken. In both, the most important thing is for the crew to systematically get the blower off the apparatus and into operation in coordination with the initial attack line (fig. 9–4). Any department can determine for itself how that is accomplished by experimenting with different deployment scenarios and using whatever response configuration is standard.

Fig. 9–4 PPA Evolution 1 assigns tasks for the first-in engine crew for fast attack. PPA Evolution 2 details tasks for the first-in engine crew when one crew member is needed up for a supply line. Courtesy Kriss Garcia

Both these evolutions have a crew of four members, each assigned to one of four specific areas of responsibility: (1) officer; (2) nozzle; (3) blower/hydrant; and (4) driver/operator. In both evolutions the driver/operator readies the engine for pumping, makes appropriate hose connections, and gets water to the attack line at the correct pressure.

PPA with fast attack

Evolution 1 deals with a typical, fast-attack PPA operation at a fire in a small dwelling (figs. 9–5 and 9–6). The duties for each crew member are as follows:

1. **Officer.** This individual takes command, makes the initial report and assignments, makes an initial size-up, and decides to go in on fast attack and not lay a water supply line from the hydrant. On leaving the apparatus, the officer grabs an appropriate tool and surveys the building, then makes or improves on a horizontal exhaust opening at an appropriate location. The officer determines when it is safe to begin pressurization and notifies crew members to direct the airstream into the entry opening. The officer then returns to the crew and supervises the interior operation while assisting with fire attack.

2. **Nozzle firefighter.** This crew member pulls the appropriate hoseline to the entry point of the building, readies it for entry, and assists with the blower.

3. **Blower/hydrant firefighter.** This crew member takes the blower and a forcible entry tool from the apparatus, extends the handle, and wheels the blower to the point of entry.

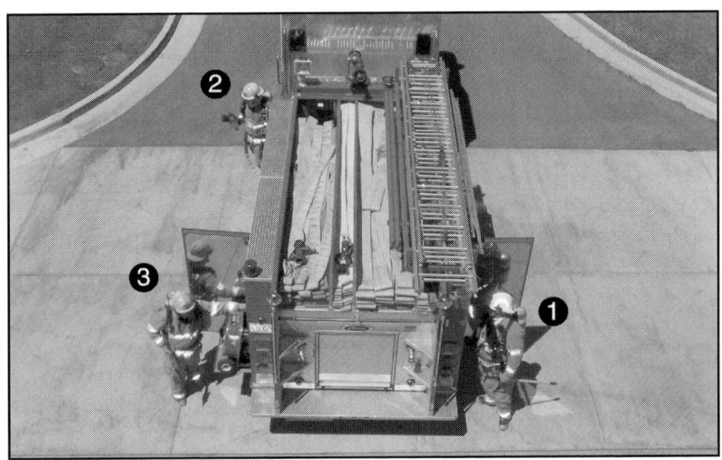

Fig. 9–5 Figures 9–5 and 9–6 illustrate Evolution 1. At the engine: the officer (1) takes one or more tools to make or improve the exhaust opening; the nozzle firefighter (2) pulls the attack hoseline; and the blower/hydrant firefighter (3) takes the blower off the engine. Courtesy Kriss Garcia

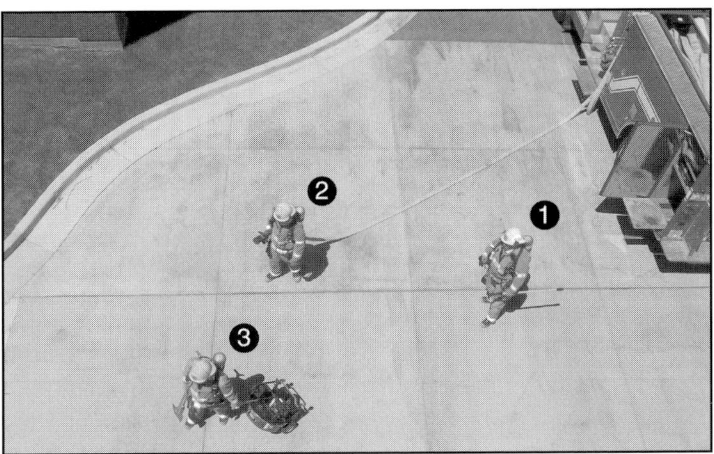

Fig. 9–6 After leaving the engine in PPA Evolution 1: the officer (1) surveys the outside of the building, makes or improves an exhaust opening, and alerts the crew when completed; the nozzle firefighter (2) carries the attack line to the building entrance and prepares for fire attack; the blower/hydrant firefighter (3) starts the blower at the entrance with the airstream directed away from the entrance. When the officer gives the okay, the airstream is directed inside, and the blower/hydrant firefighter backs up the nozzle firefighter for the fire attack. Courtesy Kriss Garcia

PPA with water supply

Evolution 2 includes PPA in a typical fire attack on a fire where the officer has determined that an immediate water supply line to the pumper is required (figs. 9–7 and 9–8). The duties for each crew member are as follows:

1. **Officer.** This individual takes command, makes the initial size-up, and decides when a water supply is needed. The blower/hydrant firefighter is directed to take a hydrant. Upon arriving at the fire building, the officer makes the initial report and assignments. On leaving the apparatus, he takes the blower and appropriate tool(s). The officer positions the blower near the point of entry, then surveys the building and makes or improves on a horizontal exhaust opening at the proper location. When the officer determines it is safe to begin pressurization, the crew members are notified to direct the airstream into the ventilation opening. The officer then returns to the crew to supervise the interior operation and assist with attack.

2. **Nozzle firefighter.** With the driver/operator, this crew member ensures that the water supply line is connected to the engine. He pulls the appropriate hoseline to the entry point of the building, prepares for entry, and assists with the blower.

3. **Blower/hydrant firefighter.** This crew member makes connections at the hydrant and then joins the attack crew.

Fig. 9–7 Figures 9–7 and 9–8 illustrate Evolution 2. The blower/hydrant firefighter connects the supply line at the hydrant. The officer takes appropriate tools to make or improve an exhaust opening and then takes the blower from the compartment. The nozzle firefighter pulls the attack line. Courtesy Kriss Garcia

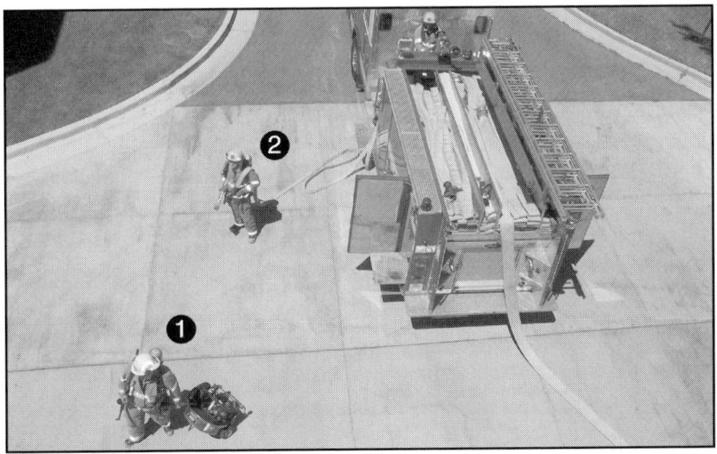

Fig. 9–8 After leaving the engine: the officer (1) wheels the blower to the building entrance, surveys the building, makes or improves an exhaust opening, and then returns to the entrance to help with fire attack; the nozzle firefighter (2) carries the attack line to the building entrance, helps with the blower, and prepares for fire attack; the blower/hydrant firefighter, not pictured here, connects to the hydrant and then helps with fire attack. Courtesy Kriss Garcia

Discussion

These two evolutions require that all crew members understand what their individual responsibilities are and what the objective is—to provide a systematic and coordinated initial fire attack. In both evolutions, crew members lay out the attack hoseline and position the blower. In both, the officer surveys the outside of the building and ensures an exhaust opening in the vicinity of the fire. The blower should be started as soon as it is delivered to the entry opening, with the airstream directed away from the entry. After the exhaust opening is made and at the order of the officer, the airstream from the blower is directed into the ventilation opening, and attack begins.

Making an exhaust opening may involve opening a door, breaking out a window, or enlarging an opening already made by the fire. Firefighters must be aware that operating a blower in an area that is on fire with no exhaust opening may increase fire intensity.

When making or enlarging the exhaust opening, the officer should keep to the side of the opening as much as possible. As hot gases exit through the exhaust opening, they move from a higher pressure and oxygen-deficient atmosphere inside the building that keeps them from expanding and igniting. These hot gases may ignite in the lower air pressure and fresh air outside. Also, since these gases can exit with force, firefighters should take precautions to protect nearby exposures, if necessary.

In addition to making an exhaust opening, the officer surveys the outside of the building. This is done to ensure there are no victims awaiting rescue at windows or other openings that may become an exhaust opening when the blower is directed into the building.

When the attack line is ready and the officer gives the command, the blower can be directed into the ventilation opening to seal the entrance with pressurized air. If the entrance is not completely sealed, pressurization may be reduced, but it will still be effective. Tests by the authors show that positioning the blower as close as 4 ft will still provide ample pressurization.

After allowing 30 seconds for the blower to begin clearing smoke and heat from the building interior, crews can advance the attack line. The reduced heat, smoke, and improved visibility enable a swift, efficient attack on the seat of the fire. The resulting steam and products of combustion are forced away from the attack crew and exhausted to the outside.

When the fire has been knocked down, the blower continues to improve the interior environment in the structure, making it easier to determine the fire cause and begin overhaul. After knocking down the fire, firefighters should immediately begin an aggressive overhaul (fig. 9–9). As previously mentioned, the rule of thumb is "Open early and open often." Since positive pressure can overcome signs of hidden fire by reducing heat and smoke, before firefighters leave the scene, blowers must be shut down and areas affected by the fire must be thoroughly checked for any hidden fire.

Fig. 9–9 Aggressive overhaul must be emphasized in training for PPA. Courtesy TCVFD

Training issues

It was mentioned earlier that these evolutions can be adapted to fit the needs and staffing levels of any department. The most important thing to keep in mind is that PPA does not require four crew members to work, but rather it requires four jobs to be performed. It just happens that these four jobs fit well with a four-firefighter crew, but if necessary, these four jobs can be accomplished with fewer people. Some departments may find it beneficial to involve two companies in PPA. They should keep in mind, though, that this introduces many potential variables. These could include arrival times, initial actions of the first-in company (as dictated by the situation), variations in staffing, equipment, and other fireground resources, and the potential for differences in training.

One aspect of PPA cannot be modified. Under no circumstances can the blower be left at the apparatus when everyone leaves to begin the initial fire attack. Experience has proven that a firefighter will not return to the apparatus to get a blower once fire attack has begun. This must be emphasized in any training.

Several other training issues also warrant mentioning. Officers and firefighters must learn that these evolutions require that they carry the appropriate equipment and tools to the fire building to perform their assigned jobs (fig. 9–10). The officer should make an adequate exhaust opening and then return to the primary duty of supervising the crew.

Fig. 9–10 When they were captains, the authors would buy pizza for their .crews if a crew member caught them at a fire without carrying a tool. Courtesy Kriss Garcia

It has been the authors' experience that if training does not continue on a regular basis, the blowers will begin to be left on the apparatus, and their benefits will not be fully realized. Without continued reinforcement, first-arriving crews will once again make the initial attack on the fire without the benefit of ventilation, thereby reducing the chances for victims in the fire area to survive. The second-arriving apparatus then gets assigned to ventilation, and a coordinated attack is not possible, because the fire attack has started without ventilation.

It has also been the authors' experience that with continual training and use of the blowers, crews will recognize how much the blowers help them to safely and more effectively do their jobs. The blowers become a standard piece of firefighting equipment, put into service by the first-arriving apparatus, because those firefighters now want the advantage that the blowers provide them. After crews have operated in an environment improved by positive pressure, they soon become a highly effective firefighting force able to provide for their own initial attack line and ventilation.

Training benefits

The result of using these two PPA evolutions or a department's adaptation of them is that the first-arriving fire apparatus becomes a self-contained firefighting unit (fig. 9–11). This unit is capable of accomplishing all of the following:

- Supplying water for the initial attack
- Advancing the initial attack line into the structure
- Beginning search-and-rescue operations with increased visibility and increased victim survival
- Providing effective ventilation that assists in victim survival, rescue operations, and finding the seat of the fire

Fig. 9–11 Providing companies with a blower and training in PPA evolutions will make each first-arriving engine a self-contained firefighting unit capable of fire attack, search and rescue, ventilation, and providing water supply. Courtesy Ray Schelble

The next chapter has answers to the 20 most common questions students ask the authors in classes they teach. The authors call these the "20 Burning Questions" about positive pressure.

Reference

1. *Webster's Third New International Dictionary of the English Language, Unabridged*. 1993. Springfield, MA: Merriam-Webster.

10 A Deeper Understanding

When the authors started teaching the theory and practice of PPA more than a decade and a half ago, the ideas were based on their practical experience and theoretical assumptions formed from logical analysis of basic physics and fire behavior. Fifteen years ago PPA was a new concept, and the supporting data for PPV timed with initial fire attack was just not available.

At each class across the country, the ideas were met with understandable skepticism and probing questions from students, especially those who had strong backgrounds in traditional ventilation methods. They fielded a lot of questions and comments from people ranging from the curious to the passionate. The authors were able to answer many, but they had to go back and dig up the answers to others. They read, talked with knowledgeable people in many fields, and engaged in long discussions with crews who were using PPA. They set up blowers in different scenarios, including countless training fires. They also had the luxury of being chief officers in a fire department that had long ago thrown its full support behind PPA, which has been perhaps the best education of all.

Some things have not changed. The authors still get a lot of questions in PPA classes, but they are now much better prepared with answers. The following is a compilation of answers to the 20 most frequently asked questions over these years. In the authors' presentations, they call them the "20 Burning Questions."

Twenty Burning Questions about Positive Pressure Attack

Much of the information in these questions is a review of information that has been discussed at length in other parts of this book. But there is still enough new information and some more detail to make them worth reading, even if readers feel they have a good understanding. If there are any questions at this point, chances are the answers will be found here.

1. Will blowing fresh air into a structure involved in fire push fire throughout the building?

No, because for the blowers to spread fire, there has to be air pushing into an area of lower pressure, which occurs as it exits to the outside. Noticeable air movement only occurs near the ventilation point and any exhaust points. Throughout the interior of the building, air movement is minimal.

One of the big, new office buildings or department stores is an example of what it is like inside a building under positive pressure. When a person first walks through the front entrance door, he feels some strong air movement to the outside through the opening (fig. 10–1). Once inside, though, any air movement as people enter and exit is not noticeable. Plants and other items show very little, if any, movement (fig. 10–2). People conduct their business in a pleasant, relaxing environment.

This is done by pressurizing the interior of the building and has the same effect as pressurizing the inside of a fire building. Raising the interior pressure slightly above the outside atmospheric pressure creates an airflow or passive shifting of environments that keeps cold, heat, dust, pollutants, and insects out of the store as people move in and out.

Fig. 10–1 Figures 10–1 and 10–2 demonstrate the air movement characteristics of positive pressure. Author Reinhard Kauffmann shows that air movement to the outside is very noticeable at the entrance door to this pressurized office building. Courtesy Kriss Garcia

Fig. 10–2 In the lobby only 10 ft from the entrance, even with the outside door open, the smoke generator and palm fronds show no appreciable air movement to the outside. Courtesy Kriss Garcia

As demonstrated through tests by the authors as well as tests conducted by AMCA, Inc., pressurization provided by commercial air-handling units or positive pressure blowers causes a passive shift in an interior environment. This shift occurs from a high-pressure area to a low-pressure area. Field tests show that when a blower is positioned outside a normal pedestrian door, air movement can be expected approximately 10 ft inside the structure, and within about 4 ft at the exit or exhaust point. Laboratory and field tests support the finding that appreciable air movement or velocity is limited to a building's points of entrance and exit. As stated earlier, a minimal increase in interior pressure, as low as 0.1 to 0.2 psi above outside air pressure, is all that is needed to maintain an effective airflow to the outside.

The answer, then, is obvious. Positioning of positive pressure blowers at the attack entrance early in the fire in coordination with the interior attack ensures the fire is confined to the area between the main body of fire and the ventilation point. The increased pressure actually holds fire extension at bay. At the same time it passively replaces the lethal heat and toxic products of combustion with an interior atmosphere in which firefighters can see and victims can survive. These profound benefits take place without creating air currents in the interior that spread fire throughout the building.

2. Will PPA push fire throughout voids in the building?

No, it will not.

One of the anecdotal arguments against positive pressure methods has assumed almost urban legend status. This legend states that pressurization while a fire is actively burning causes fire to be pushed into voids and is the culprit of numerous rekindles and massive fire extension. This is not the case.

Pressurization methods do change what firefighters have become used to in structure fire behavior. But rather than pushing the fire into voids, the impression that positive pressure is the problem stems more from its effectiveness (fig. 10–3).

The authors have studied numerous rekindle fires involving positive pressure. They also have conducted field tests where fires were deliberately started in the void space, and crews were instructed to determine whether the fire was completely extinguished. It became obvious that positive pressure does not push fire through voids. However, its effectiveness at clearing out heat and products of combustion does create conditions that can make it harder to detect hidden fire (fig. 10–4).

Fig. 10–3 Pressurization while a fire is actively burning will not push fire into voids in a building. As with any ventilation method, thorough and timely overhaul is essential. Courtesy Dan Walker

Fig. 10–4 This ceiling was completely opened up to check for fire extension into the attic space. Blowers effectively eliminate smoke and heat from a building interior, but they can also overcome signs of hidden fire. To check for hidden fire, firefighters should turn off the blowers for a minimum of 10 minutes and then thoroughly reexamine the area. Courtesy TCVFD

Twenty years ago, crews would overhaul, look for signs of hidden fire, and feel for heat to judge whether the fire was completely out. The good news is, firefighters can still depend on these senses in the same way they did 20 years ago—but not while a blower is impacting the interior atmosphere.

Positive pressure does its job well. A blower does a great job of removing heat and smoke, including the indicators firefighters depend on late in the fire to ensure complete extinguishment. In short, a blower removes drift smoke from the area as well as transient heat that indicates additional overhaul needs to be done. At this point, firefighters should turn off the blower. The heat, drift smoke, and other indicators that a hidden area needs to be opened up will become more apparent.

When considering whether additional overhaul is necessary, firefighters should remember that a blower modifies everything they have previously learned in nonpressurized attack. Adjustments have to be made. They cannot count on the absence of heat or drift smoke to determine that the fire is completely out while the blower is operating. There are two ways to help counter this changed fire behavior:

- Firefighters can aggressively overhaul all areas where fire may have traveled. As a general guideline, they should open up all spaces where fire has made contact (fig. 10–5).

- When the fire is believed to be completely extinguished, the authors recommend thoroughly reexamining the area after the blower has been turned off for a minimum of 10 minutes. As always, judgment will be the best indicator.

Firefighters must remember to turn off the blower and overhaul thoroughly!

Fig. 10–5 Thorough and timely overhaul is extremely important when using positive pressure. Firefighters should remember, "Open early and open often!" Courtesy Michael Harp

3. Will PPA spread fire through voids in a balloon frame–constructed building?

No, it will not.

Balloon frame, Type V, wood frame construction is characterized by wall studs that commonly extend from the foundation to the attic, often multiple stories, without fire stops. The balloon space in the wall, also referred to as the stud space, creates extensive void areas in all the outside walls. This presents great concern for firefighters, who know that any fire on a lower floor that finds an opening or breaks through a wall has the potential to run the wall space and quickly spread to the upper floors and attic (fig. 10–6).

Fig. 10–6 Walls in balloon construction create an open path for fire travel from the basement, seen in the photo, to the attic. Therefore, fire in the basement of a balloon-constructed structure is likely to have traveled into the attic. Crews must evaluate the status of the attic as they make entrance. Courtesy Ray Schelble

In addition to the AMCA tests referred to in question 1 that conclusively demonstrate that velocity or air movement occurs only near the ventilation and exhaust points, the authors have conducted live fire field tests on this as well. During these tests, identical balloon-constructed structures were modified in an attempt to push fire into nonfire stopped stud, joist, and rafter spaces.

The details of one of the authors' tests will be given here. In a single-story building, a 3-in. hole saw was used to make five holes, one in each stud space, through the interior wall covering into the noninsulated void space of the outside wall (fig. 10–7). These were made within 12 in. below the ceiling and 12 in. above the floor. The test fires consisted of five pallets piled on overstuffed furniture that was against the walls with the holes. The rooms measured approximately 12 ft by 16 ft, with a ceiling height of 8 ft. There was one 3-ft by 4-ft window in the test room. The room was closed off from outside air but left open to the rest of the building to allow enough air for the fire to burn. The roof had gable and soffit vents installed that were consistent with normal ventilation code requirements.

Fig. 10–7 Over the years, fires in acquired buildings have played a big part in helping the authors understand PPA. To test how pressurization would affect fire spread in balloon construction, the authors bored a 3-in. hole through the wall before starting the fire. Courtesy Alex Groves Photography

First test. This was conducted with holes only in the bottom of the wall near the floor. The fire was allowed to burn on its own until conditions in the room reached a state of imminent flashover as indicated by fire behavior near the ceiling. At this point, the blower was started at the front door, which was approximately 15 ft from the entrance to the test room. To ensure consistency and to track the thermal movement, test temperatures in the attic and center of the room stud space were recorded using a remote reading thermometer.

The blower was allowed to run for five minutes prior to fire attack to see if it would push fire into the balloon space (fig. 10–8). Measured at a point between 5 ft and 6 ft up the wall, the temperature in the stud space rose from 73°F to 86°F. The temperature in the attic rose from 79°F to 86°F. During and following extinguishment, the temperatures in the stud space lowered from 86°F to 75°F. During and following extinguishment, the temperatures in the attic lowered from 89°F to 75°F. On examining the stud space and attic area following the first test burn, there were no signs of fire extension.

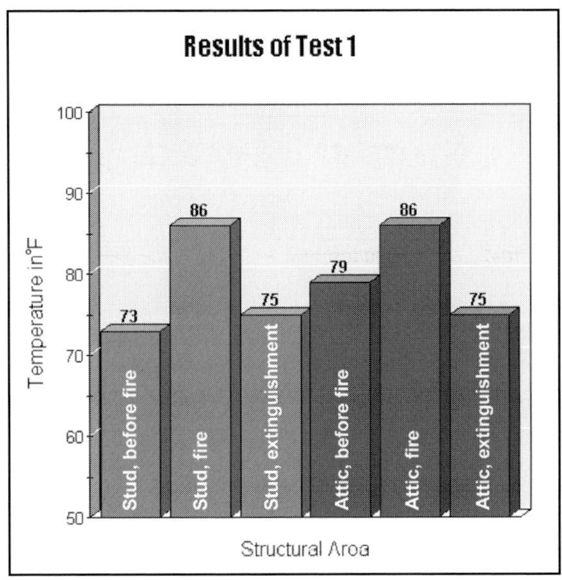

Fig. 10–8 The first test was conducted with holes in the wall near the floor. The fire was allowed to burn until it appeared to reach flashover stage, then the blower was run for five minutes to try to force fire into the wall space.

Second test. For the second test, only the top holes were uncovered (fig. 10–9). When the room reached similar flashover conditions, the blower was started at the front door, again approximately 15 ft from the entrance to the test room. During the test, temperatures were recorded in the attic and in the stud space approximately halfway between the ceiling and floor. The blower was allowed to run for five minutes prior to fire attack to again see if fire could be pushed into the stud space in the wall.

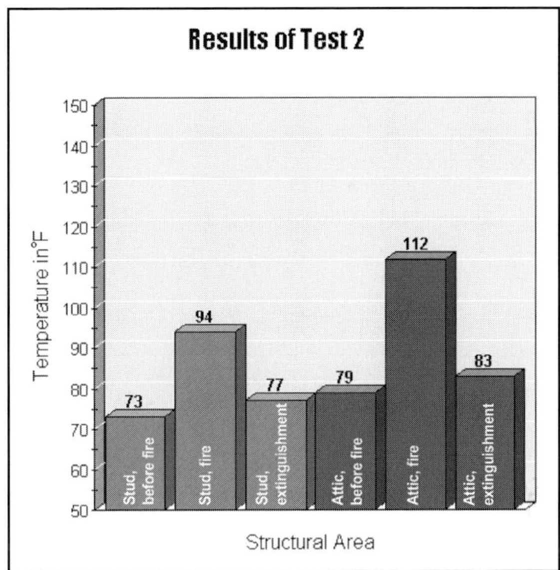

Fig. 10–9 The second test involved holes in the top of the wall.

The temperature in the stud space rose from 73°F to 94°F. The temperature in the attic rose from 79°F to 112°F. During and following extinguishment, the temperatures in the stud space decreased from 94°F to 77°F. During and following extinguishment, the temperatures in the attic decreased from 112°F to 83°F. On examination, the stud space had burn marks evident, but the marks did not extend into the attic area.

Third test. This was conducted under the same circumstances as the second, except that the attic space, which was left undisturbed during the original tests, was opened. This was done to simulate a conventional vertical ventilation hole. The fire was allowed to burn for five minutes before a vertical ventilation hole was opened in the roof (fig. 10–10). Within one minute of opening the ventilation hole, and without the blower operating, the attic temperatures rose to 187°F. At seven minutes the fire attack began, and temperatures in the attic fell to 164°F. On examining the stud space and attic area following the third test, burn marks were evident in both.

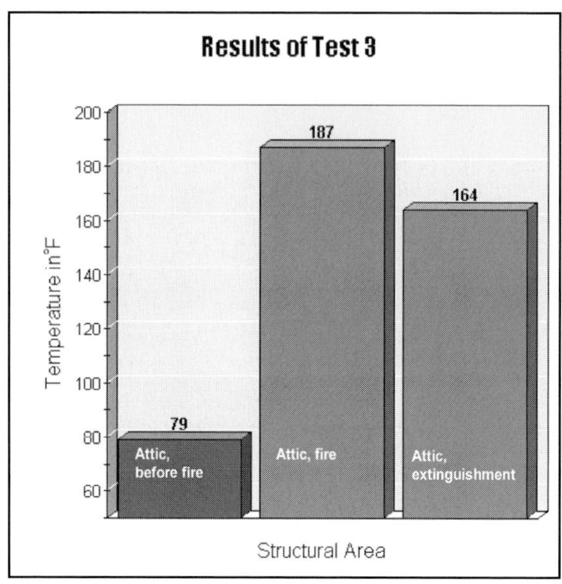

Fig. 10–10 For the third test, a vertical ventilation hole was opened in the roof after five minutes of burn time.

Conclusion. In the first two tests, a deliberately extended period of pressurization did not force fire through the intentionally breached interior walls into the stud or attic spaces (fig. 10–11). In the third test, the rooftop hole created a negative pressure space in the attic that allowed fire to extend into the stud space and attic.

In fires in most conventional dwellings of balloon frame construction, the voids are not breached to the extent they were in these test conditions. When fire gets into the balloon space in the walls, the very worst thing to do is to open the roof. This would create a negative pressure space that encourages fire spread, even without blowers pressurizing the interior of the building. Aggressive overhaul is always of utmost importance in these buildings.

Experience has proven that pressurization has no impact on fire spread into these voids. Not using blowers with an aggressive fire attack deprives initial attack crews of many benefits. This includes protecting firefighters and victims from lethal environments made more so by the disturbance to the thermal balance. Again, using PPA decreases the time for rescue, decreases the interior temperature, and improves interior conditions to the point where victim survival is more likely. It also decreases the exposure to toxic gases and smoke, and it does not push fire into these voids.

Fig. 10–11 Regardless of how pressurization may make a fire look from the outside, the risk of spreading fire to uninvolved areas in a building, even in balloon frame construction, is not high. Courtesy Rowe Harrison

4. Why does PPA sometimes make the fire appear to intensify at the exhaust point?

The majority of the fires have self-vented to some extent by the time fire crews arrive and start PPA. If not, a crew member breaks out a window, opens a door, or otherwise creates an exhaust opening. As pressurization begins to force heat and products of combustion out of the exhaust opening, it often looks as though the fire is intensifying, even though it is not.

As heat and products of combustion are forced from the structure through the exhaust opening, they come in contact with fresh air outside. At the point where the temperature, atmospheric pressure, and oxygen reach optimum levels to support free burning, these products of combustion can ignite (fig. 10–12). This most often occurs as they exit through the exhaust opening.

Fig. 10–12 Outside, at the point where temperature, atmospheric pressure, and oxygen reach optimum levels to support free burning, exhausted products of combustion will ignite. Courtesy Alex Groves Photography

During several controlled live fire field tests, the authors have observed and recorded interior conditions. This was done to demonstrate that what appears from the outside to be a mini-conflagration is actually just the exhausting of explosive interior heat and gases from the rapidly clearing and cooling interior. When interior crews are contacted in these situations, they consistently report that the interior is rapidly clearing and that they are making rapid progress to the seat of the fire.

Firefighters actually have to experience this condition to appreciate what happens. No matter how many times the authors explained to their crews what was happening, most of them had to see it from inside and outside to believe it. The biggest skeptics were convinced after they went in to attack training fires in tests and in acquired buildings while video cameras recorded outside fire conditions. After they extinguished the fire they were shown what was happening outside the building while they were experiencing cool temperatures and improved visibility inside. It made believers out of even the most skeptical.

5. Does PPA require additional staffing to accomplish the fire attack?

No, it does not.

With the proper equipment, practice, and discipline, a company can accomplish an interior attack utilizing PPA in the same amount of time as a company that does an initial fire attack without ventilation.

In addition to the responsibilities of sizing up and supervising the incident, the company officer has to take an active role in assisting with the exhaust opening. Chapter 9 discussed in detail evolutions that incorporate PPA, so the procedures and assignments will be mentioned here only briefly.

When a four-firefighter company arrives on scene, the officer has the responsibility to make or increase the effectiveness of an exhaust opening. With a three-firefighter company, the officer wheels the blower near the attack entrance as they make their initial size-up of the situation, and then he makes or enlarges an exhaust opening.

For PPA to work effectively, the first-in crew must always take the blower from the apparatus on their initial trip to the fire building (fig. 10–13). Experience has shown that firefighters find it very difficult to leave the fire building to return to the apparatus to get a blower. When this happens, they will most often proceed with fire attack and radio the next in company to initiate ventilation. Only training and discipline will ensure that first-in crews take advantage of the safety and effectiveness offered by PPA.

Having the first-arriving crew able to initiate PPA frees later-arriving companies from having to support the operation with ventilation. Later-arriving companies are now able to help with the initial attack, and more personnel are available to perform timely rescue operations, salvage, and fire control. With PPA, one company can accomplish fire attack coordinated with ventilation by using the same number of firefighters and in the same amount of time that a crew can attack a fire without PPA. Done this way, the fire attack and rescue operation are far more effective and do not create a lethal environment for the victims.

Fig. 10–13 For PPA to be an effective tactic for a fire department, first-in firefighters must always take the blower from the apparatus on their initial trip to the fire building. Courtesy Kriss Garcia

6. Does PPA delay rescue and fire control?

No, it does not.

It takes no additional time for the firefighters to wheel the blower with them as they walk toward the fire building than it does to walk without it. The key is to have the first-arriving crew take the blower to the building on their first trip from the apparatus. Again, if someone has to return to the apparatus to get a blower, crews will most often start attack without ventilation. This approach does not remove the toxic interior atmosphere and unnecessarily subjects victims to lethal conditions.

Assignment of this task has to be made routine and consistent. Assigning each crew member a specific task ensures that everything that needs to be accomplished will be accomplished.

No company should attempt an interior attack without ensuring an adequate water supply. The first-arriving officer may decide that a fast attack is the appropriate tactic. During this evolution, the nozzle firefighter pulls the initial attack line to the structure. The hydrant/blower firefighter removes the blower from the apparatus and carries tools while wheeling the blower toward the structure. The officer makes or improves the exhaust opening while doing the initial size-up of the structure. When the first-in company must take a hydrant, the duties are the same, except the officer takes the blower to the point of entry before completing the size-up and ensuring an adequate exhaust opening.

With ventilation in place, firefighters have the benefit of working in an environment that is being aggressively ventilated. Interior crews utilizing PPA can actually see their surroundings. They are able to move around far more easily, usually by walking upright instead of crawling around on all fours searching in the hot, dark, lethal environment. Some rooms can be searched in seconds simply by looking in a doorway instead of several minutes crawling in an environment that is not conducive to survival. Timely ventilation protects victims awaiting rescue, and firefighters can more effectively search for viable victims and make a rapid advance to the seat of the fire (fig. 10–14).

Fig. 10–14 Positive pressure offers flexibility to adapt to changing fireground conditions. To clear the interior faster, these firefighters put a second blower in a window to supplement pressurization after the fire was knocked down. Courtesy Richard Moseley

7. What if firefighters suspect there may be trapped victims between the fire and exhaust opening?

Unless victims are actually standing at the exhaust opening, firefighters should definitely use PPA.

By far the best chance trapped victims have for survival is for the fire attack crews to initiate ventilation immediately. This will allow firefighters to find them, and they will not be subjected to further distress from the interruption of the thermal balance caused by water application prior to ventilation.

Laboratory and live fire tests clearly demonstrate that during PPA, if victims are within about 32 in. of the floor, the temperatures they are exposed to drop almost immediately once pressurization is initiated. It does not matter where they are in relation to the fire. In tests conducted with temperature measuring instruments, victims as close as 1 meter (m) to the exhaust opening benefited from the cooling effect. Visual observations of these same conditions also demonstrate that PPA keeps heat from radiating back on the victim. It also limits the products of combustion to a point where survival is more likely than if ventilation is delayed.

One dramatic test of PPA was conducted by Chief Mark Yates of the Cornwall, England, Fire Brigade (fig. 10–15). He used recording thermocouples to document room temperatures and used mannequins with temperature-sensitive outer linings. The mannequins were initially located below the window that served as an exhaust point. They then were moved closer to the fire in successive burns until they were 1 m from the main pile of burning combustibles. Tests in each location were conducted using PPA and using fire attack without ventilation. The fires were allowed to burn to the point of flashover or a ceiling temperature of 600°F. In all cases, the temperatures on the mannequins dropped following initiation of PPA. Several observations were made as a result of these tests.[1]

Fig. 10–15 Mark Yates of the Cornwall, England, Fire Brigade studied the effects of PPA on mannequins in various locations related to the exhaust point. He also authored this study on implementation strategies and usage of PPV. Courtesy Mark Yates

Tests with the mannequins located closest to the fire. With PPA, the temperature on the mannequins dropped in every test. Without ventilation, temperatures on the mannequins increased to nonsurvivable levels.

Visual observations of mannequins. With PPA, the mannequins remained visible, and the atmosphere near the floor around them stayed clear the whole time. Without ventilation, extinguishment disturbed the thermal balance and forced a thick, choking cloud of smoke down onto the mannequins. The mannequins did not remain visible when water was applied prior to adequate ventilation. A person who cannot see through the atmosphere cannot survive breathing it.

Soot and stains on walls. With PPA, except for some damage to upper walls in the fire areas, soot stains in areas away from the fire, including areas between the fire and the ventilation opening, were minimal (figs. 10–16 and 10–17). In the fire attack without ventilation, however, walls and furnishings far from the fire in all directions were dark and soot stained. Stains in some areas adjacent to the burn area extended to floor level.

If firefighters could write their department's name on the wet, dark-stained interior surfaces after the fire is out, ventilation was not adequate. These stains on the walls are the same substances that victims are breathing and to which firefighters are unnecessarily exposed without PPA. Also, firefighters should recall that the introduction of water into a superheated environment reduces the time required for inflicting burns on skin and in the respiratory tract.

Fig. 10–16 Soot deposits on walls and belongings in figures 10–16 and 10–17 show evidence of how PPA can reduce negative impacts from heat and products of combustion. This fire attack used PPA. Notice the soot stains on the walls. In fire attack without early ventilation, soot stains will often extend all the way to the floor throughout a building. Courtesy TCVFD

Fig. 10–17 With PPA, a rapidly cleared interior limits soot to ceilings and upper walls near the involved area. The absence of heavy soot stains farther down these walls means reduced exposure to heat and products of combustion, which greatly improves chances for victim survival. Courtesy Dan Walker

PPA still provides the most effective alternative for any trapped victims, as long as they are not standing at a window or other opening that could potentially become an exhaust opening.

8. What is the best location for the ventilation opening and the exhaust opening?

The ventilation opening is where the blower is placed. This can be virtually any opening in the building that lends itself to positioning a blower. However, for safety's sake, it must be the opening that crews are using to access the building (fig. 10–18). It can be thought of as the blower pushing the first crews into the structure.

Positioning a blower at the backs of the first crews into the building allows them to take maximum advantage as the interior atmosphere clears. Sometimes they can even walk through the building, literally following behind the line of smoke and heat as pressurization clears it out. Crews can see where they are going, check for conditions that could be hazardous to their health and safety, and efficiently search for victims as they make their way to the location of the fire.

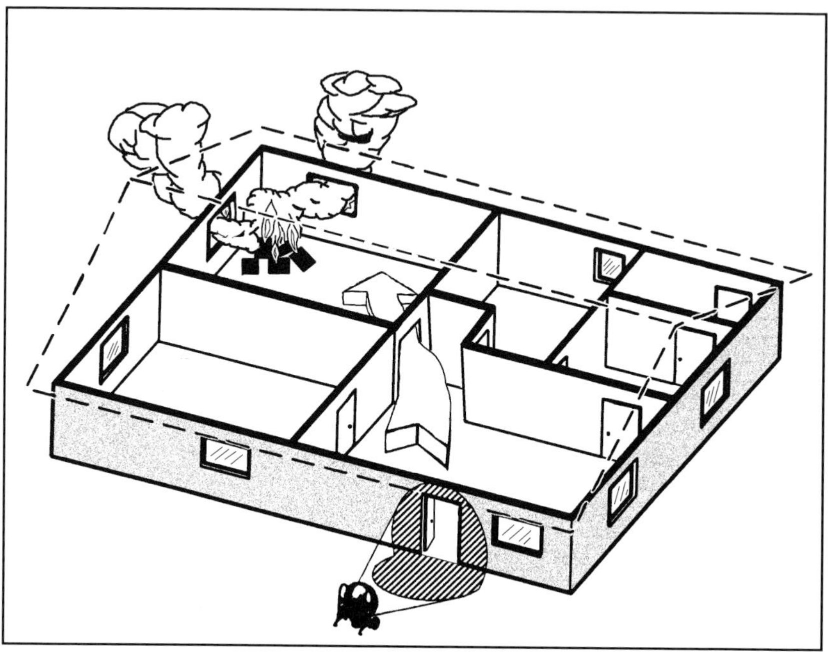

Fig. 10–18 The ventilation opening for PPA must be in the opening crews are using for access. Ideally, the exhaust opening should be as close as possible to the area of heaviest fire involvement. If the area of the fire is not obvious, the closest estimate will provide benefit if it is not separated from the fire area. Courtesy Tempest Technology, Inc.

Having a blower in any other opening without one in the access point risks creating an exhaust opening through the access, which could subject crews to exiting heat and smoke. After the fire is knocked down, pressurization can be supplemented by positioning additional blowers in other openings, such as windows. But the first blower should always be ventilating through the access point, and it should stay there.

The exhaust opening is where heat and products of combustion are forced out by the elevated interior pressure. It is best located as close as possible to the area of heaviest fire involvement. More often than not, any fire with life-threatening conditions has self-vented through an opening nearby. All that needs to be done in this case is to enlarge it, clear out any obstructions, or simply leave it alone and notify command. Sometimes there is little indication of the exact location of the fire, and in these situations firefighters can knock out a window or make an opening as close to the fire as they can estimate. This will still provide significant benefits. If available, a thermal imaging device can be a valuable tool in locating the fire.

Some places are better than others, but putting one just about anywhere will provide some benefit. Taking a "best guess" on the side of the building opposite the attack entrance is the safest approach to making an exhaust opening. Essentially, there is no bad way to get products of combustion out of the structure.

The authors have done dozens of tests where the exhaust opening was made in the "wrong" location and followed by a timely attack. On no occasion were they able to encourage fire extension as a result of ventilation pushing the fire toward the exhaust opening. Part of the reason is the drop in interior surface temperatures that occurs as pressurization introduces fresh, cool air. As stated in chapter 1, for every decrease of 18°F, the speed of the chemical reaction leading to combustibility decreases 50%. Even if the exhaust opening is in a less-than-optimum location, pressurization is still decreasing the interior temperatures, which decreases combustibility. In effect, pressurized ventilation inhibits the heat side of the fire tetrahedron by removing the superheated atmosphere.

During these tests, crews soon became aware that there was probably a better location for the exhaust opening because of the small quantities of exhausting flame and smoke. To remedy the situation, they made a second or even a third opening in better locations (fig. 10–19). With the new generation of high-volume shrouded blowers on the market today, having three to five reasonably sized (15 ft^2) openings will still yield acceptable results. If it later appears that too many exhaust openings have been made trying to find the optimum location, interior crews can simply shut interior room doors to close off ineffective exhaust openings. For more information on exhaust openings, see question 11.

Fig. 10–19 In this training exercise, crews initially made the exhaust opening the window on the far right. Because of a barrier that blocked the airflow, it did not achieve the desired result. They then broke out the window to the left of the first window, which had the same problem. Finally, breaking out the third one on the far left made the ventilation work. Courtesy TCVFD

9. At what point in the operation should the blower begin pressurizing the structure?

As companies become more comfortable with PPA, they begin to place the blower earlier and earlier in the operation. For optimum PPA, the airstream of the blower should be directed into the structure about 30 seconds prior to the attack crew's entrance. Firefighters should keep in mind that the crew member assigned to the exhaust opening always checks the building exterior as well. After ensuring an adequate exhaust opening, this crew member notifies the attack crew to direct the blower into the structure. The fire attack should begin after approximately 30 seconds. The authors and others have found that attack must be started within five to seven minutes after turning the airstream into a building, or conditions could deteriorate. It is important to have a charged attack line ready.

Fire/smoke showing. Generally, if fire or smoke or both are showing on arrival, the fire has self-vented, no victims are at windows, and fire attack can start within three minutes, the airstream from the blower can be directed into the structure as soon as possible (fig. 10–20). In this case, free burning is already taking place, and the fire has enough oxygen to burn. At the very least, pressurization will limit fire spread by confining it between the area of greatest fire involvement and the exhaust opening. If a fire has already self-vented, it is almost certain to have done so in the most appropriate location for an exhaust opening. Pressurizing the structure will simply ensure that the fire and products of combustion remain confined to the area between where the fire started and where it is exhausting.

Fig. 10–20 Generally, if fire or smoke or both are showing on arrival, the fire has self-vented, there are no victims at windows, and fire attack will start within three minutes, the airstream from the blower can be directed into the ventilation opening as soon as possible. Courtesy SLCFD

Little or nothing showing. If little or no smoke is showing outside and firefighters suspect there is a fire inside, they should not direct the blower's airstream into the structure until an exhaust opening has been made following proper procedures and crews are prepared to make entry (fig. 10–21). The crew member assigned to the exhaust opening should look for places where smoke is exhausting outside, especially in soffit areas and around windows or doors. Finding these traces of smoke will indicate where to make an exhaust opening near enough to the seat of the fire for effective ventilation. Corners of the building generally afford the best vantage point for these observations.

Fig. 10–21 If little or no smoke is showing and firefighters suspect a fire inside, the blower's airstream should not be directed into the structure until an exhaust opening has been made and crews are prepared to make entry. Courtesy Richard Moseley

10. Is it dangerous to put the blower too close to the seat of the fire?

No, this is not a problem.

The authors have done numerous live fire burns in acquired structures and test burn facilities where the fire was started adjacent to where the attack was going to be initiated. In spite of numerous serious attempts, they have never been able to "blow the fire" throughout the structure (fig. 10–22). After the building is pressurized, the fire intensifies for a short period of time and then starts to decrease in size as the atmosphere around it begins to cool.

Fig. 10–22 Over the years, the authors have deliberately set up fires right next to the ventilation point, such as during this 1994 test as recorded in this video frame. To date, they have not succeeded in "pulling" or "pushing" fire into other areas of the building. Courtesy SLCFD

Fire cannot magically transport itself from one area to another. It needs something combustible there to help it. There also have to be temperatures high enough to support combustion. As previously stated, a small drop in temperature substantially decreases combustibility. Very seldom is an area 100% full of combustibles, and without heated combustibles to carry the fire from one point to another, it will run out of fuel. All the heat the blowers move away from the fire serves to lower the temperatures of surrounding combustibles and to decrease the size of the fire.

Practically speaking, if the only place to start the attack is at the area of fire involvement, hoselines will be quickly applying water at the same time. As the products of combustion are quickly removed from the area by pressurization, so is any steam that is created by the water used to attack the fire.

11. Can a crew make an exhaust opening in the wrong location?

No, there are no bad locations for letting smoke and heat out of a building, only some that are more effective than others.

When setting up PPA, two basic scenarios are possible:

1. The ventilation opening, exhaust opening, and fire will be connected in the same space, and ventilation will occur.

2. One or two of the three will be separated by a closed door or other barrier that does not allow the flow of air.

Scenario 1. In the first scenario, all three are connected in the same space. Even if a crew member takes out a window in another room that is adjacent to the fire room, enough heat and products of combustion will exit to produce a favorable result.

In multiple tests both in training facilities and in acquired buildings, the authors have created exhaust openings in adjacent rooms to try to pull or push the fire into other areas of the building. In the absence of a room filled with 100% combustibles, the decrease in heat caused by the blowers simply does not allow the fire to communicate to other areas. When firefighters are not sure which window is closest to the main body of fire, they should make a reasonable judgment of the closest one based on the best information available. They can always try another window if the first one does not work.

Firefighters should remember that the object is not to limit the size of the exhaust opening to enhance pressurization inside the building, but rather to ensure the fire has a path to lower pressure outside. For this reason, the authors advise that exhaust openings ideally should be a minimum of two times the size of the ventilation opening. Firefighters should keep in mind, however, that positive pressure will still work effectively under far less than "ideal" conditions. The authors have achieved consistent, positive results with openings as large as four times the size of the ventilation point.

But this was not always the case. When the fire service was first experimenting with positive pressure to assist with ventilation, the recommended size of the exhaust opening was three-fourths to one and one-half times the size of the ventilation opening. The 5,000-cfm blowers and fans that were in use were not designed to take advantage of entrained air like modern blowers (fig. 10–23). The 20,000-cfm blowers in use today can provide effective air movement through several exhaust openings without exceeding their capacity (fig. 10–24).

Fig. 10–23 When the fire service first started experimenting with positive pressure, the recommended size of the exhaust opening was three-fourths to one and one-half times the size of the ventilation opening. Courtesy Kriss Garcia

Fig. 10–24 The 20,000-cfm blowers in use today can provide effective air movement through several exhaust openings without exceeding their capacity. The 5,000-cfm blowers and fans pictured in figure 10-23 were not designed to achieve airflow required. Courtesy Tempest Technology, Inc.

In a pressurized building with several exhaust openings, the exhausting products of combustion may not look as dramatic. However, the total amount of heat and products of combustion being removed will be similar to the amount removed if only one window near the seat of the fire was opened. The volume of the blower is just being split into several exhaust streams instead of one.

If the multiple exhaust openings do not accomplish the desired result, firefighters can close interior doors and passages to areas that do not need pressurization. This will isolate the necessary exhaust openings and works well to increase pressurization by making the affected area smaller as overhaul continues.

Scenario 2. In the second scenario, either the ventilation point, exhaust point, or the area involved in fire is in a separate space or spaces inside the building. This means that airflow is blocked by a physical barrier such as a wall or closed door; for example, a closed door in a bedroom on fire would separate the bedroom from the living space. A blower at the entrance to the dwelling or apartment will pressurize other interior areas but will not appreciably ventilate the fire until the bedroom door is opened. These types of conditions are not common but are certainly possible.

Ventilation that is not working properly will exhibit telltale signs. Interior indications include no or very little clearing of the interior, and increased heat if the fire is burning in the space being pressurized. As discussed earlier, pressurizing a fire area that does not have an exhaust opening can intensify the fire. Exterior indications of ineffective ventilation include nothing or very little being exhausted from the exhaust opening.

When ventilation is not working, there are two options: searching for connected space by breaking windows, and turning off the blowers. The first option, searching for connected space, requires assigning a firefighter to the outside of the building near the fire area. This may be the firefighter originally assigned to the exhaust opening or someone from a later-arriving crew. While openings in spaces separated from the fire will not ventilate the fire, they should not hurt the fire attack effort as they will help clear the interior in the uninvolved portion of the building. Searching for an exhaust opening in this way must be done in a timely manner. This is done for two reasons: to provide ventilation for the attack crew and fire victims; and, if a blower is operating, because of the risk of intensifying the fire if it does not have an exhaust opening.

The second option is to turn the blowers off or redirect the airstream away from the ventilation opening. PPA is flexible enough that ventilation can be started or stopped as necessary to adapt to changing conditions. Fire attack crews may decide it is best to enter even though ventilation is not working. Once inside, they may find a closed door or other reason why ventilation will not work. After the obstruction is cleared, these crews must judge whether to restart pressurization. Because of safety issues, all interior crews must retreat outside the attack entrance (ventilation point) before positive pressure can be started.

In both cases, communication is essential between those assigned to fire attack, the crew member ensuring the exhaust opening, and the crew member operating the blower. The ability to deal with problems like these requires ongoing training and maintaining flexibility on the fireground. There are no bad ways to remove heat and products of combustion from a structure, just some that are more effective than others.

12. What about the CO introduced by the gasoline-powered blowers?

Modern gasoline-powered blowers are designed to entrain the air around them to increase their output volume. Unfortunately, this also entrains CO from the exhaust of the gasoline engine that powers them.

Whenever these blowers are used, minimal amounts of CO are introduced into the building (fig. 10–25). Blowers manufactured after 1995 have overhead valve engines and introduce an average of 50 ppm to 100 ppm of CO into an average 2,000-ft^2 dwelling.

The top priority of any initial fire attack is to rescue victims who are overcome by high levels of products of combustion or who are trapped by fire. This is followed closely by incident stabilization and property conservation. The CO levels in a structure fire range from 800 ppm to 1,200 ppm. When attacking a dwelling fire, dropping the CO concentration from these lethal levels to a survivable level of about 100 ppm should be considered a tactical success. Firefighters should keep in mind that this applies to a well-involved room and contents fire or a larger structure fire that threatens life and property.

Effects of Various CO Levels

Carbon Monoxide Levels in PPM	Resulting Conditions/Effects on Humans
50	Permissible exposure level for 8 hours (OSHA).
200	Possible mild frontal headache in 2 to 3 hours.
400	Frontal headache and nausea after 1 to 2 hours. Occipital after 2½ to 3½ hours.
800	Headache, dizziness, and nausea in 45 minutes. Collapse and possible death in 2 hours.
1,600	Headache, dizziness, and nausea in 20 minutes. Collapse and death in 1 hour.
3,200	Headache and dizziness in 5 to 10 minutes. Unconsciousness and danger of death in 30 minutes.
6,400	Headache and dizziness in 1 to 2 minutes. Unconsciousness and danger of death in 10 to 15 minutes.
12,800	Immediate effects—unconsciousness. Danger of death in 1 to 3 minutes.

Fig. 10–25 Minimal amounts of CO are introduced by gas-powered blowers. This is not an important factor during initial attack of a working fire, but is of greater importance at small incidents. Source: American Industrial Hygiene Association

For a small fire such as food on the stove, the blowers can be used to remove smoke from the building. However, low levels of CO and gasoline fumes will be introduced into the building. On these occasions, the authors advocate removing any smoke with the blowers and then turning them off and clearing the structure of CO. Operating high-volume electric blowers is an option in these fire calls that are not life threatening (fig. 10–26). However, horizontal ventilation using open windows, especially on the windward and leeward sides, will usually take care of minor situations. HVAC systems are another option.

To examine ways to decrease CO levels, the authors evaluated several exhaust extensions. These 6-ft- to 8-ft-long tubes made of metal or rubber connect to the exhaust on the gasoline engine to pipe the exhaust a distance away. They found that these extensions provide a negligible benefit and that they can create a safety hazard for crews on the scene.

Fig. 10–26 When introducing small amounts of CO is not acceptable, high-volume electric blowers are an option. Courtesy Kriss Garcia

One of the goals when using PPA at structure fires is simply to lower CO levels from lethal levels to survivable levels, not to eliminate CO completely. At best, these extensions only lowered CO levels by less than 5%. Positioning metal and high temperature rubber extension tubes 90° windward or leeward resulted in negligible differences in the levels of CO introduced into the structure. The best results using an extension tube came from positioning the extension directly behind the blower, but it decreased the CO introduced in the building by only a few percentage points. Lengthening the tubes beyond manufacturers' lengths creates problems with engine operation.

Metal extensions become red hot. They can burn materials they come in contact with and can injure firefighters working on the scene. High-temperature rubber hoses burn through within a short period of time when the hose is bent.

In the final analysis, exhaust extensions do not provide any significant reduction of CO inside the building and add an additional 6-ft- to 8-ft-long, high-temperature hazard in the path of working firefighters. They add time to deployment of the blower and, in the authors' opinions, are not worth the trouble.

13. When should firefighters use PPA?

Firefighters should consider using PPA in any free-burning structure fire, subject to the precautions in chapter 6 (fig. 10–27). This gives victims a chance to survive and makes the interior environment clear enough so crews can operate quickly and effectively to save life and property. Also, for any situation involving products of combustion in places where they are not wanted, firefighters should consider utilizing pressurization to influence the interior environment.

Fig. 10–27 Firefighters should consider using PPA in any free-burning structure fire, subject to certain precautions. Courtesy Richard Moseley

After teaching several hundred classes to thousands of students, the authors have never heard afterward from any department or firefighter saying that what they were taught was untrue or did not work. They hear repeatedly how well PPA works and that it is unbelievable what a difference it makes in the overall effectiveness and safety of structure fires.

The authors have heard anecdotal accounts where pressurization has been blamed for serious problems, including the injury and death of firefighters. However, to date they have never substantiated one problem when crews used PPA according to proper procedures put forth in this book.

14. What about using PPA to fight attic fires?

PPA is a very appropriate tactic to fight attic fires in dwellings and businesses. Since people normally do not live in attic spaces, usually the only life risk attic fires present is to firefighters. PPA works very well to keep firefighters as safe as possible.

Setting up a blower and pressurizing the living area, merchandise area, or work area before crews enter will confine the fire to the attic space. It should also clear the interior environment and keep it that way (fig. 10–28). Interior crews operating in a pressurized area under an attic fire have better visibility for their initial search and will be better able to watch out for their own safety. If the space below is not heavily charged with smoke when interior crews start operations, no exhaust opening should be made. This maintains the highest possible pressure in the living area.

Assigning crews to a roof that has an attic fire burning below has to be considered high risk, so operations should be done from below as much as possible. Crews in the pressurized space below can make small inspection holes into the attic with a pike pole to determine the extent of fire involvement. The pressurization should prevent fire and products of combustion from extending downward through these holes. With the attic as intact as it was on arrival, small amounts of water applied through the inspection holes while operating in as safe a position as possible will often stop the forward progress of the fire and could even knock it down. With pressurization holding the fire in check, other areas in adjoining occupancies can also be checked in the same manner—from the safe, pressurized area below the fire.

Fig. 10–28 Pressurizing the living area of a structure will keep it clear and confine the fire to the attic. Courtesy TCVFD

Some in the fire service advocate applying water on an attic fire from ladders on the gable ends or from elevated platforms, both of which will work. The authors have learned that most of the time, crews going inside to perform search and rescue can apply water to the attic area faster and more effectively than crews can ladder a roof, pull a hoseline, gain access, and apply water (fig. 10–29). Interior crews take hoselines to support their search and rescue and should be checking for attic involvement as they advance into the building. While making inspection holes, it takes only a few extra seconds to open a nozzle into the involved attic space if conditions warrant. Water expands 1,700 times when it turns to steam, which makes it far more effective when using it in a confined attic space. Applying water into the attic while performing search-and-rescue operations below in a building that has been pressurized is a safe and effective practice.

Fig. 10–29 The plywood installed above this ceiling would be a serious setback for a roof crew attempting vertical ventilation. Courtesy TCVFD

After the living area is searched and water has been applied to the fire area, holes large enough to continue with overhaul can then be made from below if conditions indicate that it is safe to do so. Firefighters should look until clean, unburned structural members are found. After the attic fire is blacked out, keeping the area pressurized and opening horizontal exhaust points will remove heat and contaminants outside the building. This will make interior overhaul operations much cleaner and more effective. With aggressive overhaul through the ceiling below, a large percentage of attic fires can be handled without cutting into the roof.

In some situations it may be necessary to cut a hole in the roof for overhaul access. Firefighters should do this only after the fire has been knocked down and after making sure structural elements are safe.

Again, people usually do not live in attic spaces. The goal on attic fires is to never have any firefighter lose a life fighting a fire when hazards can be minimized through proper application of PPA.

15. What about using PPA to fight basement fires?

A stubborn basement fire is undoubtedly one of the most difficult challenges firefighters will face. In contrast to attic fires, basement fires may require aggressive search and rescue along with fire attack. Basement environments are usually difficult to clear of lethal products of combustion, and the possibility of rescuing a live victim from the basement may be low.

The two major problems with fires below grade are lack of access and difficulty with ventilation, which combine to increase the levels of heat and products of combustion. The question, then, is how to overcome these problems. The answer, as should be expected, is PPA.

Basement fires present a much different challenge for PPA than fires above grade. The natural tendency of the heat and smoke is to rise. A blower positioned on the floor above the basement must overcome this tendency and redirect the heat and products of combustion out an exhaust opening. It may require running the blower for a few moments longer before crews can safely advance downstairs.

Blower placement may call for some creativity. The standard location outside the first-floor door where crews make entry is ideal. Keeping all upstairs doors and windows closed will not only ensure pressurization in the basement, it will also help protect the upstairs from upward fire extension. However, if the upstairs is also heavily charged with smoke, exhaust openings may be necessary to help with search and rescue and to check for fire extension. Otherwise, opening a window near the entrance to the basement may be all that is necessary to keep the main floor of the building clear (fig. 10–30).

To supplement pressurization, it may be necessary to take a second blower inside and direct the airstream into the basement door opening. Multiple blowers in an in-line configuration can also help boost pressurization.

While there will always be an entrance to use as a ventilation opening, a problem can arise when trying to create a suitable exhaust opening. Due to limited windows, making an exhaust opening close to the fire may not be an option. But any opening will still provide ventilation. A second entrance to the outside could serve as an exhaust opening, but firefighters must make sure nobody uses it for access until the fire is knocked down and the interior cools.

Basement windows are often small, so several may need to be used to provide adequate exhaust openings. Firefighters should make the first exhaust opening as close to the fire as is practical. They should open additional windows as necessary, keeping in mind that at some point each additional open window may somewhat reduce the effectiveness of the pressurization.

Fig. 10–30 The exhaust opening in this basement fire was a window next to the ventilation opening, shown here in a still frame from a 1996 video. By pressurizing through the back door and leaving the upstairs closed up, the basement window becomes the exhaust opening.

16. What if a room, basement, or building has only one opening?

A door is a very necessary feature of every room. Windows, however, are not.

For a fire in a space that has only one door and no windows, pressurization is still an option to clear heat and smoke. Positive pressure can still play a role, although PPA has only a limited application. In this case the entry point, the exhaust point, and the ventilation point in the fire area must all be the same opening. Pressurizing an area that does not have an exhaust opening while the fire is free burning can create hazardous conditions for crews and victims.

For PPA to be successful, the opening must be large enough to remove high volumes of smoke or heat before crews enter. A standard doorway is not big enough. Unless the situation is right, blowers should not be used until after the fire has been knocked down. As stated previously, an adequate exhaust opening is just as important as the ventilation opening. Without adequate exhaust, crews run the risk of intensifying the fire, much like in a convection oven.

This environment also provides greatly reduced chances for victim survival due to the difficulties of removing heat and products of combustion. This is one situation where blacking out the fire before entry may be the only option, and positive pressure is adapted and used afterward to clear the interior.

When the ventilation point must also be the exhaust point, the blower must be placed differently from other positive pressure operations. The blower should be on the floor, close to the opening, but should still allow use of the door. About 12 in. to 18 in. outside the door is ideal. This pressurizes the room through the bottom of the doorway while allowing an exhaust point in the same doorway above the blower. While only about one-third as effective as PPV with a 12-ft^2 exhaust opening near the fire area, ventilating in this manner is still an option that is quicker and more effective than any other method in this situation.

Some products of combustion will reenter the area through the entrained airstream from the blower. This can largely be prevented and the operation made more effective if additional blowers are added. After placing the first blower, firefighters can add a second blower directed to blow across the opening at a 90° angle to the initial blower and about 6 ft away from it. They should tilt the second blower upward so it blows across the top of the opening, and on the windward side if it opens to the outside. This configuration moves the products of combustion away and helps keep them from becoming entrained in the airstream of the first blower.

If the doorway is located inside a structure, this operation can be enhanced by implementing PPV in the part of the building into which the room is exhausting. Firefighters can create a ventilation point near the doorway, add a third blower positioned the recommended 8 ft to 10 ft outside a ventilation point on the outside of the building, and create an exhaust point. The first blower pressurizes the area to be ventilated. The second blower removes heat and products of combustion as they exhaust through the door, and the third pressurizes the building to remove them to the outside (figs. 10–31 and 10–32).

Fig. 10–31 When there is only one opening in a space, positioning a blower to pressurize only the bottom one-half of the entrance allows the top to become the exhaust opening. The airstream of a second blower can then be directed at a right angle across the top of the opening to clear contaminants as they exhaust. The third blower clears exhausted products of combustion to the outside. Courtesy Tempest Technology, Inc.

Fig. 10–32 Ventilating a space that only has one opening is not as efficient as with separate ventilation and exhaust openings. The second blower will help remove much of the exiting products of combustion, but some will still become entrained in the airstream of the pressurizing blower. Courtesy Kriss Garcia

17. What are some potential problems with PPA?

In spite of all its good points, PPA can present problems in a few situations. These problems include exhaust opening, backdraft, noise, hidden fire, and lack of understanding and practice. Chapter 6 has a more complete discussion of the problems listed below and must be studied before PPA is attempted.

Exhaust opening. Although it is not often a problem, the first and potentially most serious possible problem is the exhaust opening. Its potential for spewing intense heat, smoke, and active flame when the blower begins pressurizing gives this the potential for being dangerous if procedures are not followed (fig. 10–33). Any window or door other than the ventilation opening has the potential to become an exhaust opening once the blower is started. People must be kept clear of the exhaust opening when PPA is started. PPA should not be used if there is a victim standing at a window until the victim can be rescued. Also, steps should be taken to protect adjacent buildings or objects located within about 6 ft of the exhaust opening.

Fig. 10–33 The potential for intense heat, smoke, and active flame to forcefully vent from the exhaust opening when the blower begins pressurizing can be dangerous if precautions are not observed. Courtesy TCVFD

Crews entering a structure before pressurization has been initiated face the possibility of being caught standing up between the fire and an exhaust opening when the blowers begin operation. It is important to always begin pressurization before crews enter or know where crews are before starting.

Backdraft. A second potential problem is when a backdraft situation exists. According to accepted practices, a structure exhibiting signs of backdraft must be vented at a high point before it is opened up low. Attempting PPA could trigger an explosion, with potentially disastrous consequences.

Noise. The third and most prevalent problem is the noise from the gasoline-powered blowers, which can either be a problem or a benefit. This is more of a problem near the blower, since the noise diminishes as crews advance into the structure. The noise can actually benefit interior crews, as they can hear when ventilation is working. Another good point for interior crews is that in the event they become lost or disoriented, they can follow the sound of the blower to locate an attack line or a route to get out (fig. 10–34).

Fig. 10–34 Noise from the blower can be a problem or a benefit. Courtesy Dan Walker

Noise does somewhat compromise the verbal communication of crews positioned outside near the blower, but it should not be of much concern to the IC, who should not be that close to the incident.

Hidden fire. A fourth potential problem with PPA is that following knockdown and during overhaul, a blower can create a deceptively clear atmosphere. Crews must realize that the blower will overcome heat and smoke that would normally indicate hidden fire. If interior conditions are clear with the blower operating, it should be stopped a minimum of 10 minutes prior to doing a thorough, final evaluation of the structure (fig. 10–35). If in question, the motto still applies: "Open it up!" Many rekindles have been blamed on pressurization, when further research has shown that this is not the case. As with any method of ventilation, crews that do a complete and thorough overhaul will find any fire extension and complete the extinguishment.

Fig. 10–35 Many rekindles have been blamed on pressurization when that was not the case. Firefighters should aggressively open up areas that have been impinged by fire and turn off the blowers for at least 10 minutes before giving the area a thorough final check.

Lack of understanding and practice. This final problem with PPA may actually be the most important. On the surface it appears to be a simple operation—firefighters set up a blower and advance a hoseline. It is certainly not a complex operation, but there is far more to it than that. As with any other tactic, PPA must be fully understood and regularly practiced to be an effective, safe method for fire control.

Sometimes crews can get by for years without fully knowing all that they should know. But if they do not understand its intricacies, however few and simple, eventually they run the risk of putting their crews and civilians in danger. Although there are few conditions and situations where PPA can pose problems, they must be respected.

18. Under what conditions is PPA not recommended?

There are four situations when PPA should not be used.

The first is when firefighters or victims are standing at windows or other openings that could become an exhaust opening once the blower is started. In this situation, rescuing those individuals should be the highest priority. It may be safest and fastest to complete a ladder operation in these circumstances.

The second situation is when fire crews may be operating beyond the fire in a location that may be between the fire and an exhaust opening (fig. 10–36). The best protection against this is a coordinated attack where the IC knows where all crews are operating. Ensuring that crews do not advance ahead of the fire and put themselves between the fire and the exhaust opening is an absolute necessity. To avoid this situation, firefighters should always start ventilation before entering. For post-knockdown PPV, firefighters should leave the structure before starting the blowers and then reenter through the ventilation opening.

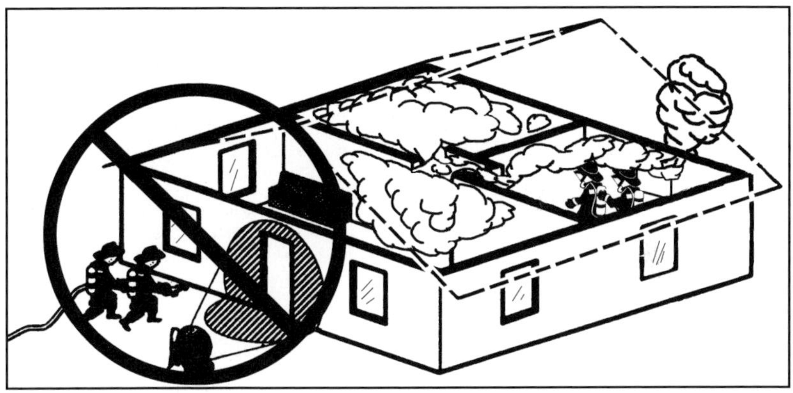

Fig. 10–36 A building should not be pressurized if firefighters may be between the fire and the exhaust opening. Graphic courtesy Tempest Technology, Inc.

The third condition is when there is combustible dust or flammable vapor. Fires in grain elevators or other occupancies that are at risk of dust explosions should be controlled utilizing a tactic that is least likely to increase suspended dust (fig. 10–37). Often this will be through the judicious application of water, usually by misting the area. Caution should be taken to not move any combustible atmosphere to points of possible ignition as the environment is displaced.

Fig. 10–37 Blowers must not be used in grain elevators or occupancies containing combustible dust. Courtesy Kriss Garcia

The movement of flammable vapors can be very unpredictable. It is recommended that before pursuing any course of action that will cause the movement of flammable vapors, firefighters should consult with hazardous materials technicians (fig. 10–38). This will help to determine correct identification and ensure vapors will not be moved into an environment that could escalate the incident or create separate problems. Explosion-proof electric blowers are available to deal with these types of situations.

Fig. 10–38 Firefighters should not use blowers on flammable liquids and flammable vapors. Courtesy Scott Freitag

The fourth situation in which PPA should not be used is when backdraft conditions exist (fig. 10–39). Any structure exhibiting signs of backdraft conditions must be ventilated at a high point before fresh air is introduced at a low point. Fires that have settled into the smoldering stage leave little hope of rescuing viable victims. Pressurization can be used after proper initial ventilation to remove heat and products of combustion.

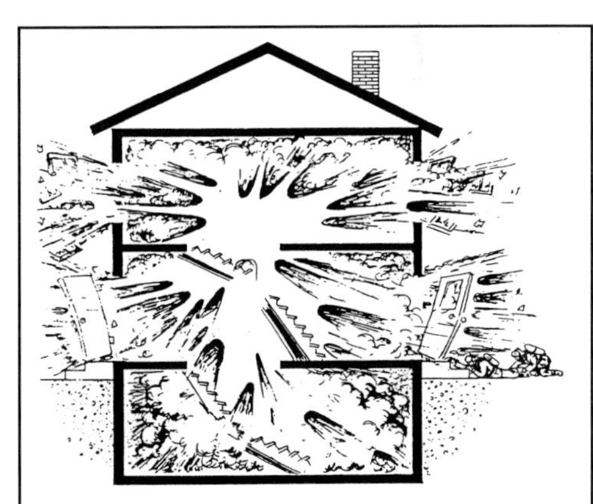

Fig. 10–39 When backdraft conditions exist, the chance of finding viable victims is very low, and improperly adding fresh air could have a violent result. Courtesy SLCFD

19. How can PPA protect exposures?

Pressurizing a building interior makes it difficult for fires to burn into it. Pressurization can provide excellent protection to buildings or other areas exposed to adjacent fires (fig. 10–40). The method works whether the exposure is an entire structure or a partitioned area in the same structure. This could be a shop in a strip mall, a dwelling with an attached garage, or a floor in a high-rise, when another part of the building is involved in fire.

Fig. 10–40 Pressurization was used as exposure protection in the white building on the left to help protect it from a fire involving the white trailer visible in the center of the picture behind the tree. The trailer contained unstable dynamite. Courtesy Martha Ellis

The higher the pressure in the interior, the more effectively it will keep fire out (fig. 10–41). For exposure protection, crews should not open any exhaust openings in the uninvolved exposure area and should keep blowers as far away from the fire as possible. Positioning the blowers on the side opposite where fire is impinging lessens the possibility that the blower will push brands or flaming debris into the exposure.

Crews still need to stand by inside any exposure, and all voids that are directly impinged on by the fire should be opened and observed. An exposure operation does not exhaust the interior environment. The exposure may be clear of smoke and products of combustion, but CO levels should still be monitored and evaluated. Sometimes crews assigned to protect exposures may have to operate in SCBA because of the CO levels, even though the interior looks clear. This is especially important if crews are operating in this environment for more than 30 minutes. Firefighters should test the area to ensure CO is within safe limits or that proper precautions are taken. SCBA should be worn at all times if the area is not being tested.

Fig. 10–41 Internal exposures, such as offices when other areas of the building are burning, should be priorities for protection. Protecting records in an office can save a business. Courtesy Ray Schelble

This use of positive pressure has had positive effects when used for wildland urban interface exposures, but many burn-over situations result in flaming debris on all sides of the exposure. When the air around a structure is full of flaming debris, firefighters should keep the exposed structure closed up. If, however, the flaming debris is limited and the firefighters need to ensure that small fires in the structure or on the roof do not spread rapidly into the structure, then the use of positive pressure may be warranted. As stated earlier, this method will not protect combustible exterior siding or roofing. In these cases, positive pressure may buy time for crews to track these areas down and extinguish them.

20. What size of building can be ventilated by a blower?

After almost a decade and a half of tests in acquired structures and of using positive pressure regularly at incidents, the authors feel safe in saying that a single 20,000-cfm blower can effectively ventilate residential or commercial buildings up to 5,000 ft^2, or 40,000 ft^3. The exiting volume is what is important.

With the same size ventilation opening and exhaust opening and the same capacity blower, the volume of air moved in a building with 1 million ft^3 of space will be the same as in a building that is 100,000 ft^3. The difference is that the larger building needs more volume moved to achieve the same effect as in the smaller building. More volume can be achieved by either using multiple blowers or a larger blower, such as an MVU.

Conclusion

PPA is a safer, more effective option over other types of ventilation, including vertical ventilation, for several reasons (fig. 10–42).

Fig. 10–42 This young man has taken a more than casual interest in Salt Lake area fire stations. He carries his gear with him, the same as his heroes. He even carries a fan with him, in order "to blow the smoke away," he tells people. Courtesy Reinhard Kauffmann

Benefits of PPA

Safety. PPA involves a simple, rapid operation that generally takes place outside and on the ground. Vertical ventilation requires using ladders to access a roof system that was not designed to support the additional weight. It requires multiple tasks including laddering, adding additional hoselines, and cutting a hole. In addition, it takes place on a roof that is subject to influences such as structural instability, weather, and changing fire conditions.

Effectiveness. PPA confines the fire by utilizing openings near the fire that the fire has either already made or openings that already exist and can be improved on rapidly. Vertical ventilation requires opening an additional path for fire spread.

Coordination. PPA incorporates ventilation with initial attack, allowing firefighters to make viable rescues and attack the fire more effectively. Vertical ventilation is time-consuming, and crews simply will not wait for it to work before entering the building to make a fire attack. This disturbs the interior environment to the extent that anyone not protected by structural firefighting gear and SCBA may not survive.

Efficiency. PPA does not require assigned crews or even extra personnel. Vertical ventilation requires assigned personnel to accomplish multiple tasks.

Flexibility. With PPA, if situations arise with which crews are not comfortable, they can make adjustments or completely stop the operation if conditions warrant. Once vertical ventilation has been established, crews can do nothing but react to the conditions the hole has created. They cannot change them or stop the process.

The next and final chapter offers some final thoughts and observations on PPA.

Reference

1. Yates, Mark. 2001. The wind of change. (Brigade Command dissertation). Fire Service College, Coventry University, 2002.

11
Final Thoughts

It is your decision where to go with what you have learned.

We have given it our best shot. We have provided objective and proven information. We have talked about methods of ventilation. We have spelled out the basic principles of positive pressure and provided detailed, step-by-step instructions on how to go about safely using PPV and PPA. We have showed you how a variety of positive pressure methods for ventilation and fire attack can be adapted for many situations. We have given you models of evolutions to help your department focus on how to best accomplish the necessary functions of PPA. We have answered some of your most pressing questions. Our attempt to educate you is nearing an end.

PPA is controversial in some circles, but it is no different from any other fireground tactic. It is not an end in and of itself. It is just one more tool that can help us accomplish what is of ultimate importance in our profession: to safeguard the lives and property of the citizens we protect and to ensure our crews return safely to the station afterward. It certainly is not appropriate in all situations, and firefighters who use it carelessly or without adequate knowledge and training can definitely make a bad situation worse.

Each fire is different. The infinite variations firefighters face at every fire provide appropriate circumstances to utilize every type of tactic at one time or another. For this reason, intelligent firefighters will not utilize tactics based exclusively on procedure, habit, tradition, or what some "guru" says. We need the ability to observe and reason when looking at how tactics can contribute to the positive outcome of an incident. Properly performed PPA definitely adds safety and effectiveness to a broad range of fire operations. However, you should resist any temptation to think of it as a "no-brainer" that can be used without considering each fire's unique circumstances.

Many have heard this or that horror story circulating of how PPA does not work or how it made this fire or that fire worse. We have heard them, too. We can honestly say that in every horror story we have researched, the cause could be traced back to errors in PPA's implementation, such as misuse of the blower or a lack of coordination. These types of errors almost always come from a lack of knowledge and training. It really is not that much different in this respect than just about any other aspect of fireground operations.

The bottom line is that done properly, PPA is a safe, effective tactic that can help you do your job if you make an earnest attempt to learn how it works, practice what you learn, and use adequate equipment. You can attack a fire quickly and effectively, which greatly increases the chances of rescuing viable victims, lessens damage to your citizens' property, and reduces the risks to your crews. If at any time PPA is not accomplishing what you feel it needs to, simply turn off the blower, and it goes away.

After teaching positive pressure methods to thousands of firefighters just like you, from large and small departments across the nation and internationally, we are confident your experiences will be similar to the success stories we have consistently heard from our students.

This book is just one step. Make a difference!

A

Why Positive Pressure Works

Reinhard Kauffmann

None of us involved in the writing of this book are scientists, and physics and chemistry were not our college majors. However, Kriss Garcia and I felt we needed to present at least a short and hopefully simple explanation of the science behind why increasing the pressure inside a burning structure with a blower moves heat and the products of combustion to an area of lesser pressure.

The Science behind Positive Pressure

The science behind positive pressure ventilation can be as easy as A, B, C, and D. This concept can be explained by the four gas laws that are commonly referred to as the A, B, C, and D of the Ideal Gas Law: Avogadro's Hypothesis, Boyle's Law, Charles' Law, and Dalton's Law. When all of these laws are considered, they form the basis of what is known as the Ideal Gas Law.

Avogadro's Hypothesis

This hypothesis states that equal volumes of gases, under identical temperature and pressure conditions, have an equal number of molecules. This number is referred to as Avogadro's Number, and it is the number of molecules in a mole of any substance—6.0221367×10^{23} molecules per mole. A mole is the atomic mass of a given substance represented as a numerical weight in grams. The important thing to remember is that equal volumes of gases have the same number of molecules, or the same number of moles, when the pressure and temperature are constant. One liter of gas A has the same number of molecules as one liter of gas B.

This can be related to the extinguishing agent of choice for firefighters throughout the centuries, namely water (H_2O). Four liters of hydrogen (H) have twice the number of molecules, or moles, as one liter of oxygen (O_2). When chemically reacted, the product is two molecules of water. Written as a chemical equation, it is: $4\ H + O_2 = 2\ H_2O$. Readers who have taken a chemistry or physics course know Avogadro's number better than they know their own phone number, and many would just as soon forget it!

Boyle's Law

Robert Boyle determined that if the temperature of a gas remains constant, the pressure of that gas is inversely proportional to the volume of the gas. In other words, if one increases the pressure of gas, one will find that the volume has decreased. Consider what happens when firefighters refill their SCBA cylinders after a fire. Air is forced into the cylinder under pressure to provide a greater amount (or mass) of air in a smaller volume (the cylinder) for a firefighter to breathe. When firefighters put on their SCBA and use the air, the high-pressure air in the cylinder moves to an area of lower pressure through a system of hoses and regulators, and then into their lungs.

Lungs also function according Boyle's Law. During inhalation, the diaphragm moves downward, and the chest expands in size. This results in the volume of the lungs increasing, and the pressure inside the lungs decreasing. Air at a higher pressure outside one's body moves into the area of lower pressure within the lungs. (One should recall that decreased pressure results in an increase of volume.) When air is exhaled, the diaphragm moves upward, and the chest decreases in size. The result is a decrease in the volume of the lungs and an increase in the pressure of the air in the lungs. Thus, the air at a higher pressure moves to the lower pressure area outside the body.

Next one should visualize the blower at the ventilation point of a structure involved in fire and an exhaust point being established near the main body of the fire. The blower forces air into the structure, which increases the interior pressure. This means that whatever gases are inside are compelled to move to lower pressures on the outside through the exhaust opening. The gases inside, of course, are the heated products of combustion. When these products of combustion move to the outside, their volume increases because they are no longer confined (compressed) inside the structure. Their pressure has been reduced. (High pressure = low volume; low pressure = high volume.) This is why one sees the large amount of smoke that is now in an area of decreased pressure billowing from the exhaust point. The process continues because the blower maintains the elevated pressure inside.

In case one is wondering, the mathematical formula for Boyle's law is:

$$P_1V_1 = P_2V_2$$

where

P represents pressure and
V represents volume.

The subscripts 1 and 2 represent the initial pressure and volume and the resulting pressure and volume, respectively.

Charles' Law

Charles' Law is similar to Gay-Lussac's Law. Charles did much of the scientific discovery, but he never bothered to write and publish his findings. When Gay-Lussac published similar discoveries, the law was named after him. But, because Gay-Lussac referred to work originally done by Charles, most people started referring to the law as Charles' Law.

Charles' Law states that at a constant pressure, the volume of a given quantity of a gas varies directly with its absolute temperature. This law is somewhat easier to understand than the other laws because of the direct relationship between temperature and volume. As the temperature increases, so does the volume of a gas, which in this case is smoke. The assumption made in describing this law as it relates to a structure fire is that as the temperature of a fire increases, the structure must have a large enough vent opening to accommodate the increase in the volume of gases in order to keep the pressure constant inside. In other words, with no outside influence from a blower, temperature increases inside a building require a correspondingly larger ventilation opening to maintain a constant pressure inside.

What this means to firefighters is that when an operating blower is directed into a fire structure, the temperature is reduced because of the introduction of cool air. It will proportionately reduce the volume of gas (smoke) being produced.

The mathematical expression for Charles' law is:

$$\text{V is proportionate to T}$$
$$V/T = \text{constant}$$
$$V_1/T_1 = V_2/T_2$$

where

V represents volume and
T represents temperature.

The subscripts 1 and 2 represent the initial volume and temperature and the resulting volume and temperature, respectively.

Gay-Lussac's Law approaches the problem from a slightly different direction. It states that if the volume of a given quantity of gas is held constant, the pressure of the gas varies directly with the absolute temperature.

The mathematical expression of Gay-Lussac's Law is:

$$\text{P is proportionate to T}$$
$$P/T = \text{constant}$$

where

P represents pressure and
T represents temperature.

It is of interest that both Charles and Gay-Lussac were involved in hot air ballooning, and their studies led them to discoveries that were beneficial to their favorite sport. They figured out that increasing the temperature of the gases used to inflate their balloons resulted in an increased volume, which made their balloons fill more quickly. Also, the first blowers developed by Mr. Dexter Coffman, founder of Tempest Technology, Inc., were created to inflate hot air balloons and were later modified for ventilation during firefighting.

Dalton's Law

The fires that firefighters encounter every day contain a multitude of different gases within the smoke that is produced. One of Dalton's contributions to science was to state that the pressure exerted by each gas in a mixture is called its partial pressure. The total pressure exerted on a container is the sum of the partial pressure of each gas in the mixture. Stated another way, the pressure of each gas that makes up smoke is the same as if it were the only gas in the structure.

$$P_T = P_A + P_B + P_C \ldots$$

where P represents pressure. The subscript T represents the total pressure of all the gases in the smoke, and the other subscripts represent the partial pressures of individual gases.

The Ideal Gas Law

This all brings us to what is called the Ideal Gas Law, or the way that gases would behave universally if they were ideal gases. Unfortunately, just like there is no ideal firefighter, there is also no such substance as an ideal gas. There are other considerations that complicate what actually constitutes an ideal gas, such as diffusion, solubility, and a host of others that go beyond the scope of this brief explanation. At any rate, the Ideal Gas Law does explain why gases move from one area to another area due to variations in pressure, temperature, and volume. To bring this back to PPV, it explains what happens when a blower forces more air into the interior of a structure than is exhausted, namely that air pressure within the space is greater than that on the outside.

Mathematically, the Ideal Gas Law is expressed as:

$$PV = nRT$$

where

P is the pressure of a gas,
V is the volume of gas,
n is the number of moles of the gas (recall Avogadro's number),
R is a constant for all gases, and
T is the temperature of the gas.

The Ideal Gas Law generally holds true for all gases, including mixtures of gases at any temperature, pressure, or volume, even though no gas is truly ideal.

Gas laws in real life

For a real-life example, one could consider the situation that a firefighter faces when responding to a house fire. We will try to explain what occurs with the smoke using the A, B, C, and D of the gas laws. Avogadro's Hypothesis tells us how many molecules there are within a measurable volume of the individual gases in the smoke the firefighter sees on arrival. The smoke that is escaping the structure through any and all openings is due to the volume of smoke being compressed as the pressure inside increases, just as Boyle described. The pressure exerted by all of the lethal gases can be measured according to what Dalton discovered. Now Charles said that if one increases the temperature, as in a structure fire, one will also get more smoke.

Unfortunately, all of this is working against the firefighter's attempts to find any survivable victims or to advance to the seat of the fire. This brings us back to Boyle's Law and the concept of using a blower to mechanically increase the interior pressure. This will move those heated products of combustion to an area of lower pressure, namely outside the structure. The blower also decreases the temperature, which will also result in a decrease in the volume of smoke.

All of this ends with the Ideal Gas Law, which as mentioned previously is only good for the ideal or perfect world and explains the phenomena observed in our structure fire only if many other variables are considered. However, it does explain the general behavior of gases or smoke.

The point of this discussion and the point of this book is to help firefighters gain a fundamental understanding of what makes positive pressure attack work, so as to utilize it as a fireground tactic.

B

Ventilation Evolution

Adopted 10/97

Ventilation Evolution #1

Positive Pressure Attack (PPA) with fast attack

Purpose: To increase the effectiveness of a fast attack on a fire by implementing PPA.

Description: The apparatus arrives at the fire scene and goes in with a fast attack using a 1-3/4 inch line supplied by tank water and PPA. Evolution begins when apparatus stops at the scene and ends when PPA is in place and the attack line is flowing an effective stream and advanced inside.

Notes: ❑ Safety and common sense take priority in all fireground activities. Use all required safety equipment, including full PPE where needed. ❑ Do not use PPA if a backdraft condition exists. ❑ Allow blowers to operate for 30 seconds before advancing the attack line inside. ❑ The entrance opening must be kept clear—firefighters standing in the entrance opening can seriously reduce the effectiveness of PPA. ❑ Apparatus operated with headlights on. ❑ Open and close gate valves slowly and carefully with two hands. ❑ Seat belts must be used when seated on apparatus. ❑ Straddling hose or hose clamps prohibited at all times.

Position	***Procedures*** *(Also see the chapters, "Fire Attack" and "Ventilation," in the SLCFD Training Supplement.)*
Officer	❑ **Assigns tasks:** Notifies crew and incoming companies of the decision to go in on a fast attack and that a later-arriving company may need to lay a supply line. After determining a backdraft condition does not exist, gives the order to initiate PPA with fast attack. Advises hydrant/fan firefighter on the placement of the fan. Informs nozzle firefighter to pull a 1-3/4 inch pre-connect attack line. ❑ **Sizes-up building:** Grabs an appropriate tool and sizes-up the building. If necessary, makes or improves on a horizontal vent at a location near the fire. ❑ **Supervises and assists:** Returns to crew and supervises PPA and the interior operation. Helps gain entry through the entrance opening and assists with the fire attack.
Nozzle	❑ **Pulls attack line:** As directed by the officer, pulls the appropriate attack line and advances it to the building entrance used for the fire attack. ❑ **Entry and attack:** If necessary, helps gain entry through the entrance opening. Begins fire attack after the blower runs for 30 seconds.
Hydrant/ Fan	❑ **Positions blower:** Secures the blower and a forcible entry tool from the apparatus. Extends the handle on the fan and wheels it to the point of entry as directed by the officer. ❑ **Operates blower:** When attack line is in position and ready, starts the blower and positions it properly in the entrance opening. Ensures the opening remains clear.
Driver/ Operator	❑ **Sets-up apparatus:** Positions apparatus. Before leaving cab, sets parking brake, engages pump, and puts apparatus into proper gear. ❑ **Safety:** Places chock under apparatus wheel, and positions safety cones to warn traffic. ❑ **Charges attack line:** Opens tank drop. Slowly charges the attack line by partially opening the discharge valve until the line is filled, then opens the valve to provide an effective stream.

Comments:

Adopted 10/97

Ventilation Evolution #2

Positive Pressure Attack (PPA), taking a hydrant

Purpose: To increase the effectiveness of a fire attack requiring water supply from a hydrant by implementing PPA.

Description: At the hydrant — The crew takes a forward lay from the hydrant with LDH or 2-1/2 inch hose following procedures in Hose Evolution #1 or Hose Evolution #2, then drives to the fire scene. At the fire scene — The crew completes the hydrant connections and attacks the fire using PPA and a 1-3/4 inch or 2-1/2 inch attack line. Evolution begins when apparatus stops at the fire scene and ends when the attack line is flowing an effective stream and advanced inside.

Notes: ❑ Safety and common sense take priority in all fireground activities. Use all required safety equipment, including full PPE where needed. ❑ Do not use PPA if a backdraft condition exists. ❑ Allow blowers to operate for 30 seconds before advancing the attack line inside. ❑ The entrance opening must always be kept clear—firefighters standing in the entrance opening can seriously reduce the effectiveness of PPA. ❑ Apparatus operated with headlights on. ❑ Open and close gate valves slowly and carefully with two hands. ❑ Seat belts must be used when seated on apparatus. ❑ Straddling hose or hose clamps prohibited at all times.

Position	Procedures *(Also see the chapters, "Fire Attack" and "Ventilation," in the SLCFD Training Supplement.)*
Officer	❑ **Assigns tasks:** *At the hydrant* — Directs crew to take a hydrant with LDH or 2-1/2 inch hose following procedures in Hose Evolution #1 or Hose Evolution #2. *At the fire scene* — After determining a backdraft condition does not exist, informs crew that PPA will be initiated and directs nozzle firefighter to pull a 1-3/4 inch or 2-1/2 inch attack line. ❑ **Positions blower:** Secures the blower and an appropriate tool from the apparatus. Extends the handle of the fan and wheels it near the point of entry ❑ **Sizes-up building:** Carrying an appropriate tool, sizes-up the building. If necessary, makes or improves on a horizontal vent at a location near the fire. ❑ **Assists with attack:** Helps gain entry through the entrance opening and starts the blower if it is not already running. When attack line is in position, assists with the fire attack.
Nozzle	❑ **Pulls attack line:** As directed by the officer, pulls the appropriate attack line and advances it to the building entrance used for the fire attack. If necessary, attaches a hose strap. ❑ **Entry and attack:** Helps gain entry through the entrance opening. Assists with blower. Begins fire attack after the blower runs for 30 seconds.
Hydrant/ Fan	❑ **Assists with attack:** After taking the hydrant following procedures in Hose Evolution #1 or Hose Evolution #2, goes to the objective with forcible entry or overhaul tools in hand.
Driver/ Operator	❑ **Sets-up apparatus:** Positions apparatus. Before leaving cab, sets parking brake, engages pump, and puts apparatus into proper gear. ❑ **Safety:** Places chock under apparatus wheel, and positions safety cones to warn traffic. If necessary, places hose clamp on the supply line, on the hydrant side of the last coupling. ❑ **Gets water into pump:** Begins operation using tank water if appropriate. Sees that the supply line is connected to the pump intake. Removes hose clamp if necessary. After the supply line is charged, slowly opens the gate into the pump intake. Changes-over from tank water to hydrant water if necessary. ❑ **Charges attack line:** Slowly charges the attack line by partially opening the discharge valve until the line is filled, then opens the valve to provide an effective stream.

Comments:

C

AMCA Standards for Blowers

The following two pages are taken from ANSI/AMCA Standard 240-96, which describes a new method adopted by the AMCA to provide certified ratings of their ventilation capacity. Previous test methods evaluated only the blowers' output in cfm and made it difficult to compare different blowers.

With positive pressure ventilation, design has a major impact on the blower's effectiveness at ventilating a structure. The major elements of a blower that affect the exhaust from the AMCA test chamber are blower diameter, blade design, shroud design, and the associated air entrainment. The amount of exiting air volume from the test chamber is directly affected by the air exchanges a blower can provide to the test area, thus measuring the relative effectiveness of appliances used in fireground positive pressure operations. The AMCA label on a blower certifies that its performance specifications have been tested and verified.

The following illustrates the testing procedure, and after that, sample test data are presented (figs. C–1 and C–2). More information on AMCA International can be found at www.amca.org.

AMCA STANDARD 240-96

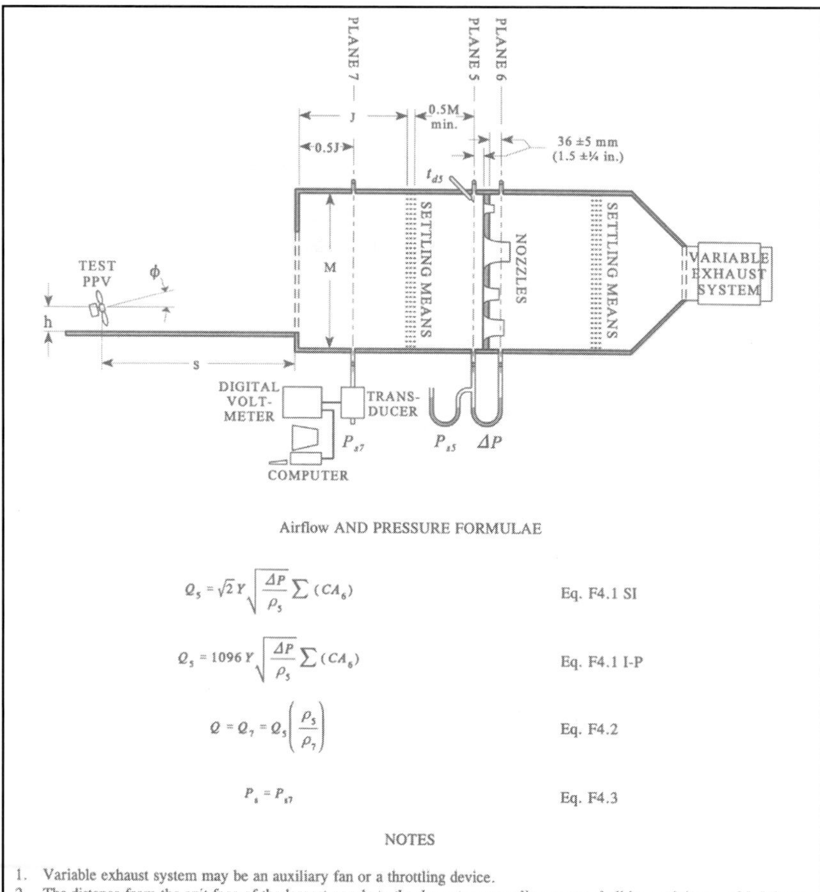

Airflow AND PRESSURE FORMULAE

$$Q_5 = \sqrt{2}\,Y\sqrt{\frac{\Delta P}{\rho_5}}\sum(CA_6)$$ Eq. F4.1 SI

$$Q_5 = 1096\,Y\sqrt{\frac{\Delta P}{\rho_5}}\sum(CA_6)$$ Eq. F4.1 I-P

$$Q = Q_7 = Q_5\left(\frac{\rho_5}{\rho_7}\right)$$ Eq. F4.2

$$P_s = P_{s7}$$ Eq. F4.3

NOTES

1. Variable exhaust system may be an auxiliary fan or a throttling device.
2. The distance from the exit face of the largest nozzle to the downstream settling means shall be a minimum of 2.5 throat diameters of the largest nozzle.
3. Dimension J shall be at least 2.0 times the PPV equivalent discharge diameter.
4. Temperature t_{d7} may be considered equal to t_{d5}.

Fig. C–1 The AMCA testing prodecure.

TEST #761-P04 **02/08/95**

AMCA Laboratory - Arlington Heights, Illinois

UNIT

Name:	Windmaker 324
Model No.:	ADF1-P24
Impeller Dia.:	0.61 m
Appurtenance(s):	Windmaker, Type D Misting Nozzles

SETUP

s = 3.05 m
ϕ = 11 °
h = 0.46 m

DATA

t_{d0} = 27.3 °C
t_{w0} = 18.5 °C
p_b = 745.4 mm Hg

i determination	N (rev/s)	t_{d5} (°C)	P_{s5} (Pa)	ΔP (Pa)	t_{d7} (°C)	P_{s7} (Pa)
1	58.32	28.9	-24.6	1461.1	28.9	-24.6
2	58.37	28.8	-20.1	1307.1	28.8	-20.1
3	58.38	28.8	-14.4	1210.3	28.8	-14.4
4	58.43	28.7	-9.9	1106.4	28.7	-9.9
5	58.43	28.4	-5.2	988.2	28.4	-5.2
6	58.48	28.2	4.7	702.4	28.2	4.7
7	58.43	28.1	9.9	557.1	28.1	9.9
8	58.50	28.0	14.7	422.2	28.0	14.7
9	58.52	27.8	19.4	338.3	27.8	19.4
10	58.52	27.8	24.6	189.5	27.8	24.6

RESULTS

i determination	ρ_0 (kg/m^3)	$P_{si} = P_{s7}$ (Pa)	$Q_i = Q_7$ (m^3/s)
1	1.1460	-24.6	12.609
2	1.1460	-20.1	11.921
3	1.1460	-14.4	11.469
4	1.1460	-9.9	10.963
5	1.1460	-5.2	10.380
6	1.1460	4.7	8.740
7	1.1460	9.9	7.779
8	1.1460	14.7	6.788
9	1.1460	19.4	6.070
10	1.1460	24.6	4.535

CURVE COEFFICIENTS

K_0 = 30.589620
K_1 = 0.318340
K_2 = -0.372880

AIRFLOW AT FREE AIR

Q_f = 9.494 m^3/s

Tested by RDK.

Fig. C–2 Sample test data.

Reference

1. ANSI/AMCA Standard 240-96. 1996. *Laboratory Method of Testing Positive Pressure Ventilators for Rating*. Arlington Heights, IL: Air Movement & Control Association.

D

Acronyms and Abbreviations

AMCA	Air Movement & Control Association International, Inc.
ANSI	American National Standards Institute
BTU	British thermal unit
cfm	cubic feet per minute
ft	foot
gpm	gallons per minute
HVAC	heating, ventilation, and air conditioning
IC	incident commander
ICS	incident command system
in.	inch
LAFD	Los Angeles City Fire Department
MVU	mobile ventilation unit
NFPA	National Fire Protection Association
NIOSH	National Institute for Occupational Safety and Health
OSB	oriented strand board
PPA	positive pressure attack
PPE	personal protective equipment
PPV	positive pressure ventilation
PVC	polyvinyl chloride
RECEO-VS	rescue, exposure, confine, extinguish, overhaul, and ventilation and salvage
SCBA	self-contained breathing apparatus
SLCFD	Salt Lake City Fire Department
TCVFD	Tooele City Volunteer Fire Department
UJNR	United States–Japan Cooperative Program in Natural Resources
VES	vent/enter/search

Bibliography

Air Movement & Control Association International, Inc. *Laboratory Method of Testing Positive Pressure Ventilators for Rating*. Arlington Heights, IL: Air Movement & Control Association, 1996.

Allmon, Cliff. "Positive Pressure Ventilation." *Fire Chief*. May 1988: 39–41.

Ames, Stephen. "Behavior of Materials in Fire." *Materials World*, 1 (2), February 1993.

Appleby, Phil. "Tactical Exchanges." *Fire Prevention Fire Engineers Journal*. April 2003: 46–47.

Beal, Kellie Ann, ed. "Thirteenth Meeting of the UJNR Panel on Fire Research and Safety March 13–20." Vols. 1 & 2. Gaithersburg, MD: National Institute of Standards and Technology, 1997.

Brannigan, Francis L. *Building Construction for the Fire Service*. Quincy, MA: National Fire Protection Association, 2000.

———. "A Call for Action by Fire Chiefs." May 2000. www.fireengineering.com.

———. "The Unexpected Killer." January 1999. www.fireengineering.com.

Brunacini, Alan V. *Fire Command*. Quincy, MA: National Fire Protection Association, 1985.

Coffman, Dexter. "Big Wind: Positive Pressure Ventilation Gains Foothold in Fire Fighting." *Industrial Fire World* (January/February) 23–28, 2000.

———. Personal correspondence to Bill Manning (editor, *Fire Engineering Magazine)*. September 25, 1999.

Dunn, Vincent. *Collapse of Burning Buildings*. Saddle Brook, NJ: PennWell Publishing, 1988.

Egan, Mark. "The Application of Scale Modeling to Assess the Effectiveness of Positive Pressure Attack in Large Single Storey Buildings." B.E. diss., University of Central Lancashire, 2004.

Everton, A.R. " PPV—Its Use and Non-Use: A Lawyer's Perspective." Seminar paper prepared for the South Western Branch of the International Fire Engineers. April 2003.

Fire Engineering. "Firefighter Dies after Falling through a Roof Following Ventilation—the NIOSH Report." May 2003.

Firetactics.com. "Positive Pressure Ventilation & Casualty Location." 2003. www.firetactics.com.

Fried, Emanuel. *Fireground Tactics*. Chicago: H. Marvin Ginn Corp., 1972.

Garcia, Kriss. "Response to August Roundtable Responses to PPV." December 1999. www.fireengineering.com.

Grimwood, Paul. "Positive Pressure Ventilation—A Review of International Research." 2002. www.firetactics.com.

———. "What the French Say about Positive Pressure Ventilation." *Fire & Rescue*. April 2003: 30.

Hall, Richard, and Barbara Adams, eds. *Essentials of Fire Fighting*. 4th ed. Stillwater: Fire Protection Publications, Oklahoma State University, 1998.

Institute for Research in Construction. "Ottawa Fire Department Puts IRC Research into Practice." 5 (2) 2000. http://irc-cnrc.gc.ca.

Karter, Michael J. Jr.,"2002 U.S. Fire Loss." *NFPA Journal*. September/October 2003: 59–63.

Kerber, Stephen, and William D. Walton. "Characterizing Positive Pressure Ventilation Using Computational Fluid Dynamics." Gaithersburg, MD: National Institute of Standards and Technology, 2003.

Klaene, Bernard J., and Russell E. Sanders. *Structural Fire Fighting*. Quincy, MA: National Fire Protection Association, 2000.

Lawson, J. Randall. "Thermal Performance and Limitations of Bunker Gear." August 1998. www.fireengineering.com.

Lawson, J. Randall, and Nora H. Jason, eds. "Firefighter Thermal Exposure Workshop: Protective Clothing, Tactics, and Fire Service PPE Training Procedures." *NIST Special Publication 911*. Gaithersburg, MD: National Institute of Standards and Technology, 1996.

McAniff, Edward P. *Strategic Concepts in Fire Fighting.* Saddle Brook, NJ: PennWell Publishing, 1974.

Mendes, Robert F. *Fighting High-Rise Building Fires.* Boston: National Fire Protection Association, 1975.

Montgomery County Fire & Rescue Services. "Managing the Consequences of a Chemical Attack—a Systematic Approach to Rescue Operations." Montgomery, MD: Montgomery County Fire & Rescue Services, December 1998.

National Fire Academy. *Strategy and Tactics for Initial Company Operations.* 1st ed. Emmitsburg, MD: National Fire Academy, 2002.

National Fire Protection Association. *Fire Protection Handbook.* 19th ed. Quincy, MA: National Fire Protection Association, 2003.

———. NFPA 1001, *Standard for Fire Fighter Professional Qualification.* Quincy, MA: National Fire Protection Association, 2000.

National Fire Service Incident Management System Consortium Model Procedures Committee. *Model Procedures Guide for High-Rise Fire Fighting.* 1st ed. Stillwater: Fire Protection Publications, Oklahoma State University, 1996.

National Institute for Occupational Safety and Health. "Fire Fighter Fatality Investigation Report 98-07." 1998. www.cdc.gov/niosh.

———. "Fire Fighter Fatality Investigation Report 98-07." 1998. www.cdc.gov/niosh.

———. "Fire Fighter Fatality Investigation Report 99F-23." 2000. www.cdc.gov/niosh.

———. "Fire Fighter Fatality Investigation Report F2001-13." 2002. www.cdc.gov/niosh.

Phoenix Fire Department. "Positive Pressure Ventilation." Phoenix Regional Standard Operating Procedures M.P.202.12D. Phoenix: City of Phoenix, 1995.

Praxair, Inc. "Carbon Monoxide." Material Safety Data Sheet. Danbury, CT: Praxair, Inc., 1999.

Prendergast, Edward J. *Building Construction Related to the Fire Service.* 2nd ed. Stillwater: Fire Protection Publications, Oklahoma State University, 1999.

Pressler, Bob. "Operations on Peaked-Roof Structures." July 1995. www.fireengineering.com.

Reneke, Paul A., et al. "A Comparison of CFAST Predictions to USCG Real-Scale Fire Tests." *Journal of Fire Protection Engineering.* 11 (1): 43–68, 2001.

Richard, Bruce. "The Routine Garage Fire." May 2002. www.fireengineering.com.

Super Vacuum Manufacturing Company, Inc. *Smoke Ventilation Training Manual.* Loveland, CO: Super Vacuum Manufacturing Company, Inc., October 1999.

Tuomisaari, Maarit. *Smoke Ventilation in Operational Fire Fighting.* Espoo, Finland: VTT Publications, 1997.

U.S. Edgewood Chemical Biological Center. *Use of Positive Pressure Ventilation Fans to Reduce the Hazards of Entering Chemically Contaminated Buildings.* Aberdeen Proving Ground, MD: 2000.

United States Fire Administration. *Fire Data Analysis Handbook.* 2nd ed. Emmitsburg, MD: United States Fire Administration, January 2004.

———. *Firefighter Fatality Retrospective Study 1990–2000.* Emmitsburg, MD: United States Fire Administration, April 2002.

———. *Risk Management Practices in the Fire Service.* Emmitsburg, MD: United States Fire Administration, 1996.

Vaari, Jukka, and Jukka Hietaniemi. *Smoke Ventilation in Operational Fire Fighting.* Espoo, Finland: VTT Publications, 2000.

Vettori, Robert L., Daniel Madrzykowski, and William Walton. "Simulation of the Dynamics of a Fire in a One-Story Restaurant—Texas February 14, 2000." Gaithersburg, MD: National Institute of Standards and Technology, 2002.

Yates, Mark. 2001. The wind of change. (Brigade Command dissertation). Fire Service College, Coventry University, 2002.